EVERYDAY LAS VEGAS

EVERYDAY LAS VEGAS

*Local
Life in a
Tourist
Town*

REX J. ROWLEY

UNIVERSITY OF NEVADA PRESS RENO & LAS VEGAS

University of Nevada Press, Reno, Nevada 89557 USA
Copyright © 2013 by University of Nevada Press
Manufactured in the United States of America
Design by Kathleen Szawiola

Portions [or Parts] of chapter 8 were included in "Religion in Sin City,"
GEO REVIEW (American Geographical Society) 102, no. 1
(2012): 76-92, and are reprinted with permission.

Library of Congress Cataloging-in-Publication Data

Rowley, Rex J., 1977–
Everyday Las Vegas : local life in a tourist town / Rex J. Rowley.
p. cm.
Includes bibliographical references and index.
ISBN 978-0-87417-905-7 (cloth : alk. paper) — ISBN 978-0-87417-906-4 (ebook)
1. Las Vegas (Nev.)—Description and travel. 2. Las Vegas (Nev.)—Social conditions
—21st century. 3. Las Vegas (Nev.)—Social life and customs—21st century.
4. Working class—Nevada—Las Vegas—Social life and customs. 5. Tourism—
Social aspects—Nevada—Las Vegas. 6. Working class—Nevada—Las Vegas—
Interviews. I. Title.
F849.L35R69 2013
979.3´135—dc23   2012037472

The paper used in this book meets the requirements of American National
Standard for Information Sciences—Permanence of Paper for Printed Library
Materials, ANSI/NISO Z39.48-1992 (R2002).
Binding materials were selected for strength and durability.

University of Nevada Press Paperback Edition, 2014
23   22   21   20   19   18   17   16   15   14
5   4   3   2

ISBN-13: 978-0-87417-945-3 (pbk. : alk. paper)

FOR RACHEL

My companion and confidante,
in all life's geographic explorations

You see that there's more to Vegas than the Strip when you see all the other lights

—Comment by a tourist from Sheffield, England,
following a "City Lights" helicopter tour

CONTENTS

ILLUSTRATIONS

MAPS

PREFACE

This book is about my city. Although I grew up in Las Vegas, it was only much later that I realized what a unique place it is. When I tell someone where I am from, I often encounter weird looks, puzzled reactions, and surprised excitement. I have come to enjoy such moments, as they reinforce the connection I feel to a city perceived as exotic and strange to people across the country and around the globe. Being from a unique place gives me a sense of pride in my own identity; because of my hometown, I feel I too am different. The guy from Las Vegas is often the one who gets attention in a crowded room.

Yet, at such times, I also think to myself: "Was my life in Las Vegas really all that different from someone living in any other American city?" As I formalized my training as a professional geographer, I realized that the answers to such a question are not so simple. Places are complex things, and take on different personalities for different people; perceptions of a place are, after all, based on the experience each person has within it. Tourist locations, such as Las Vegas, have a particularly bifurcated personality: insider residents see the place through their experiences of work, worship, play, and raising a family, whereas outsiders imagine the same location through a set of tourist perspectives. While not so pronounced, a similar pattern exists for nontourist places; visitors and passers-through will always see a place differently than longtime residents. I realized my city might be unique, but the reality of my place attachment based on insider familiarity is not unlike the experience of millions of humans in thousands of towns and cities around the world. Las Vegas is simply a place where the insider/outsider division is starkly evident.

Because of the huge media campaigns and the millions of tourists who visit the city each year, the outsider's image of Las Vegas is often the only one talked about. Roughly two million people live in the areas surrounding the Strip, but their stories often go untold, to the point that some outsiders don't believe people actually live in Las Vegas. Given the unique nature of the desert metropolis and the fact that it is the quintessential example of the insider/outsider dichotomy so common in places, I felt that that pattern and how it plays out here (and, by extension, elsewhere) needed a book-length exploration.

I set out to portray the "other side" of this bifurcated place and sought to answer slightly altered versions of the question I had been asking myself: "What is life in Las Vegas like for local residents, and how do they interact with the tourist side of the city?" "What do such interactions say about the city's personality and sense of place?" I try to answer these and related questions through an ethnographic approach that included interviewing more than a hundred residents, participating in local culture and activities, and keeping a finger on the pulse of the community through its newspapers. As a geographer, my focus is on place identity as constructed through human interactions within certain spaces. I dig deep into the stories, anecdotes, and insights of residents living in this city—especially within the spatial context of the tourist destination—to construct a perceptual map of life in this unique place that evokes its bifurcated personality. Seeing this one exemplary city in such a light will lead to other questions and answers regarding how we as humans see and interact with all our places.

Although the cover has my name on it, I could not have completed this book without the help of others. My gratitude begins and ends with my family. My wife, Rachel, has read (and reread) every word of this book. She has listened to my complaints, offered solutions to my research and writing dilemmas, and opened a willing heart and mind to the sacrifices our family would make to see this book through.

Our children—Matthew, Caleb, Laurelin, Maggie, Rex, and Sam—have been patient with their dad as he worked many late nights and postponed playing more games of catch or tag or Slapjack in order to finish his education and this book. I thank them for their smiles, hugs, and simple, yet impressive words of encouragement.

My mother and father, Rexine and Dennis Rowley, provided material support for my work, a place to live in Las Vegas while I conducted my research, and even became research assistants when I needed a photograph or two and could not make the trip back home. They have always encouraged me to believe—in example and spoken word—that I could do and be anything. I also owe gratitude to my siblings and grandparents for their friendship and heritage. I give special thanks to my late grandfathers, Richard Rowley, whose life as a writer and college professor inspired me, and Rex (Pop) McAllister, my namesake, friend, and the most ardent supporter of my chosen profession.

I thank Pete Shortridge for his guidance on the project, from inception to completion. The idea for this book came out of a lecture he gave in his cultural geography class during my second year as a graduate student. My love for place and the inner workings of how humans interact with their geography expanded through many chats with him over the decade that followed.

I also want to recognize the contributions and encouragement of colleagues: Terry Slocum, Garth Myers, So-Min Cheong, Steve Egbert, Dave McDermott, Carol Bowen, Linda Sue Warner, Jon Thayn, and Rich Waugh. Thanks to Bill Tsutsui for providing validation that this portrait of Las Vegas would be of interest to people outside my sometimes insular geography bubble.

Thank you to Matt Becker for his enthusiastic reception of this project and for seeing its potential as a book. He and the staff at the University of Nevada Press have shepherded me along a heretofore-untraveled road of professional publication. I appreciate the effort of the reviewers and editors and express gratitude to John Kostelnick and Kathleen Szawiola for their help when it came to producing the maps I have included.

Finally, I thank my fellow everyday Las Vegans, past and present, for making this city what it is and for opening their doors and minds to me and giving me something special to write about. This book is my effort to characterize the place they have built. I hope that my interpretation of their words has done justice to the ideas and impressions they shared in conversations about our place in the desert.

EVERYDAY LAS VEGAS

The Local's Las Vegas

L as Vegas is one of the world's most recognizable cities. People all over the world seem to have some notion of what this place is like. Frank Simon, a longtime Las Vegan, told of his visit deep into the "outback of the Outback" in Australia. He and a friend went into a hotel where he wrote down his address at check-in. The woman behind the desk asked him repeatedly: "You're from *the* Las Vegas?" When Frank replied in the affirmative, she disappeared into the back room and told a friend, who let out a blood-curdling shriek. Shocked, Frank listened as she explained that meeting him was "probably as close to Las Vegas as I will ever get." She asked him: "Do you know George? Do you know Brad?" Frank gave me a wry look, "you connect the dots." I laughed and replied, "as in *Ocean's Eleven* George [Clooney] and Brad [Pitt]?"

Such an episode underscores two important elements of the city's character. On the one hand, the woman's reaction stems from the "Vegas image" in popular culture. She was enamored with the glitz, glamour, and twenty-four-hour entertainment. This is the image captured in the Strip's iconic skyline and presented in movies, TV shows, and worldwide advertising campaigns, such as the award-winning "What happens here, stays here" commercials. It is the image that made the "Las Vegas brand" America's second-most popular in the Newsmaker Brands Survey in 2006, right after Google and ahead of iPod and YouTube. And pollsters expected this ranking to hold in later years.[1] Her perception of Las Vegas was propped up by tales and pictures brought back home by the more than thirty-five million visitors to the city per year for the past decade.[2] And the impressions carried by these visitors—the Strip and its giant themed hotels, glittering lights, gambling, elaborate Cirque du Soleil shows, adult entertainment, fine dining, and endless buffets—are representative of what is understood of Las Vegas by most people.

On the other hand, the derisive manner in which Frank told the story points to misconceptions about the life of residents of this city. Las Vegans commonly meet strange reactions when explaining where they are from.

Many residents can recount similar questioning responses from outsiders: "Which hotel do you live in?" "Is your father a dealer and your mother a showgirl?" "Wow! I didn't know people actually lived in Vegas. What's that like?" Las Vegans often chuckle and consider such perceptions outlandish, even unbelievable. Some respond sarcastically: "Yes, I do live in a hotel, and when my dad is off with the mob boss and my mother goes off to work in the show, I curl up under a blanket in a dark corner of the poker room." Others try to counter the misconceptions by claiming that life is as normal here as it would be anywhere. Regardless of the response, that a rejoinder is even necessary illustrates how living in Las Vegas is a unique experience. Even though Las Vegans go to work, attend church, or visit a city library, just like Americans across the country, they do all this in the shadow of a flashy tourist destination known far and wide. Some locals are directly involved in the tourist realm, through employment or lifestyle, while others have a more implicit relationship with the "Vegas image," if only because they have to affirm that they *do* live in Las Vegas. In other words, Las Vegas residents dwell in an entangled separation with the Vegas of popular imagination.

Such varied perceptions of Las Vegas epitomize the bifurcated nature of the city's personality. It is a town of insiders and outsiders, and the different perceptions are directly related to individual experience within it, whether from the local or tourist perspective. Such is the story of how humans interact with place. Even though Las Vegas may be an exaggerated case, it is nonetheless an example of how each of us sees our place—at the most basic level of an insider or an outsider—based on our own encounters in it.

Most portrayals of Las Vegas give only scant acknowledgment of the people that live in the shadow of the neon. After all, visitors to Las Vegas are not likely to focus on the blanket of lights *away* from the Strip and downtown. They probably will not visit a grocery store, see a high school football game, make an appointment with a local physician, or gamble at a neighborhood bar. And yet these everyday things are part of life for the city's two million residents. Even less attention has been given to how locals interact with the overwhelmingly powerful tourist side of their city. As a result, Las Vegas is one of the least understood famous places in the world, a modern-day terra incognita.

This book turns the tables, so to speak, on the typical portrayal of Las Vegas in order to present another image, another side of the city. This alternative perspective does not speak of "what happens here, stays here," but rather, "what stays here, happens here." It is the Las Vegas as experienced

This south-facing view on Wichita Falls Street near Centennial Parkway and Clayton Street is characteristic of the new North Las Vegas suburbs and the insider-outsider dichotomy visible on the city's landscape. Photo by author, June 2007.

by the local people and how their identity in place is subject to the insider/outsider binary so present in this city. Such a portrayal of one particular city can help us see other places in a new light.

THE INSIDER AND THE OUTSIDER

Every story has two sides. This axiom applies in cases as disparate as parents quelling a conflict between siblings, attorneys arguing the cases of plaintiff and defendant, or diplomats determining the appropriate placement of a disputed border. Geographers interested in the human relationships that exist between, or within, places, see the "two sides to every story" axiom from a spatial perspective. They ask, for example, how does an immigrant generation's view of its new home differ from that of their offspring in the second generation? How does a child, unaccustomed to the views and biases of adults, perceive and interact with space? What meaning do places of memorial hold for those connected to a disaster or incident as opposed to people who have no direct experience with that event?[3]

Realistically speaking, many more than just two perspectives exist for places and the stories of place. In fact, each and every person that encounters a place will likely have a different perception of it. This tenet certainly holds true for Las Vegas. Still, a simple dichotomy between the views of the insider and the outsider is useful to consider. Think, for example, about Lawrence, Kansas. Although to some outsiders this particular place might be just another town on the pancake-flat Great Plains, most Americans probably know it as a college town with its quirky mix of culture, art, and college sports. This outsider's view, although true, is also quite limited. To an insider, experience with and connection to the place are much deeper. Lawrence is not only a college town, but a place becoming both a retirement center and bedroom community of Kansas City, where people establish themselves and their families because of local amenities such as a quaint and vibrant downtown and a still tangible small-town feel. Finally, and most importantly, to the resident insider, Lawrence is home, containing all of the images, feelings, and attachments that title holds.

Speaking of the human experience in landscape, Robert Riley once described the insider/outsider distinction in terms of visual cues. An outsider's perception of place, he noted, is composed of elements that are entirely visual, those that are typically a source of pleasure to the person observing a landscape. Such a characterization might describe many of the thirty-five million visitors to Las Vegas who often make the obligatory walk up and down the Strip to "view" what the place has to offer. Riley further speculated that, as one spends more time in the landscape, this visual experience diminishes. The insider's (local resident's) landscape, then, becomes "internally experienced" and "far richer and more personal."[4] Similarly, Douglas Pocock has written that experience in place leads to a "characteristic bounding with internal structure and identity, such that insideness is distinguished from outsideness." He noted, for example, that some place-based novels written by persons from outside the locale are not well received by insiders.[5]

Perhaps Kent Ryden has best characterized the insider/outsider dichotomy and its relevance for Las Vegas. In his book about sense of place in Coeur d'Alene, Idaho, Ryden writes that tourists focus only on "surfaces, with things that a place contains." Visitors develop perceptions of a place based on superficial qualities rather than deeper layers of meaning. The tourist's sense of place for Las Vegas is naturally based only on the location (usually just the Strip) they visit, their landscape of experience. Ryden continues: "The people who live in that place, however, could tell that viewer of a wide variety of mythic, legendary, historical, and personal meanings overlying what may have seemed to be a largely neutral chunk of geography." The resident's experience in Las Vegas takes a measure of the tourist landscape (and some of the legends that follow it), but adds to it the experience of gambling at a so-called neighborhood casino, worshipping with a local congregation, frequenting a local park, and living within a suburban neighborhood. In this way the landscape that is invisible to the tourist, or outsider, is uncovered through lived experience of the resident insider. As Ryden has phrased it: the "unseen layer of usage, memory, and significance" experienced by an insider has the power to open up "an entire new world before [the outsider's] eyes."[6] For Ryden (and myself) the *in*visible landscape is the landscape of the *in*sider. My aim is to extract some of the "rich" and "personal" connections to place in Las Vegas, portray some of the city's "internal structure and identity," and uncover a heretofore hidden facet of the Las Vegas landscape.

UNCOVERING A SENSE OF PLACE

Such an aim evokes questions related to the human experience in space and place. What exactly do we mean when we speak of these two geographical entities? What is a sense of place? Succinct, encompassing definitions for such terms are difficult to form, but a brief discussion of how I grapple with these fuzzy and fluid concepts in the context of my work can be helpful.

Places are created out of our experience within specific *spaces*. To illustrate, imagine the first day a newcomer drives into the town that will be her home for the next several years. She tries to follow directions to the apartment she rented, but misses several turns. Later that day, she looks for a department store, failing in the attempt. The newcomer feels as if she is in a maze. Because of inexperience, she has difficulty finding her way, and her focus on specific tasks (navigating a moving truck along unknown streets and alleys) prevented her noticing the passing landscape except for street names, to which she paid particular attention. As she gets to know her new home in the days and months that follow, she becomes familiar both with the ways of getting around and the personality of the city. What began for her "as undifferentiated space [ended] as a single object-situation or place." Furthermore, her experience along the way—at a friend's house near one of the wrong turns she made the first day, or at that store she had initially intended to find—gives deeper meaning to the city and its landscape. In effect, she discovers that "when space feels thoroughly familiar to us [through experience], it has become place."[7]

My goal is to reveal the local side of Las Vegas as a distinct *place*. This is not to say that the local side of the city is completely separate from that experienced by a tourist; the local's Las Vegas, in fact, is a portion of the *space* that is the whole city but is endowed with different meaning based on the experience of those within it. Stated differently, the millions of people who visit Las Vegas each year may realize that a large population lives within the city—it would be impossible for them not to see the burgeoning suburbs as they drive or fly into town—but since tourists have not experienced life in Las Vegas, their only view of the inhabited city is as *space*. The resident, through lived experience, finds value, meaning, and identity and their city becomes a *place*. This dichotomy is not that different from the woman's experience in a new town. Her position was similar to a tourist until she had experienced and formed her own ideas of what the city is, at which point she became a local: the space became place. For my portrait of Las Vegas, I want to illuminate the city's local *sense of place,* which can be described as the feelings of identity, value, familiarity, and attachment among people in a place that gives a locale its recognizable character and personality. Places, just like people, have personalities, and it is only through experience that we can come to know what that personality is.

How does one identify the personality of a place? Such a task can be as overwhelming as defining it, largely because of its indivisible attachment to the subjective intangibles of human experience and feeling. Yet, that subjective attachment provides an answer to how sense of place may be found. We must look to methods of gathering, analyzing, and describing experiential human knowledge; we must look to the stories local residents tell about their place. I have done so for Las Vegas using three main sources.

The first is a set of interviews I conducted with residents over a three-year period between 2005 and 2008. The stories, anecdotes, impressions, and descriptions I heard in these meetings offered rich glimpses into life in the city. In all, I held lengthy conversations with one hundred Las Vegans and informal, shorter conversations with seventy-seven additional residents. Most of my interviews were conducted in confidentiality, and the interviewees' actual names are withheld by mutual consent. I employ pseudonyms to give identity to their remarks. In several cases I obtained written consent to quote civic leaders, journalists, and others whose identities strengthen the concept or argument in which they are included. In these cases, the speakers' actual names are cited in the text. Each interview is listed separately in the bibliography.

I also have clipped thousands of articles, commentaries, and columns over the past eight years from the two local daily newspapers, the *Las Vegas Sun* and *Las Vegas Review-Journal,* as well as other weekly and monthly publications. In addition, while in the field interviewing, I lived in a Las Vegas neighborhood, participated in community events, observed the actions of myself and other local residents, and recorded dozens of resulting thoughts, anecdotes, and interpretations.

Finally, I grew up in Las Vegas, and even though I have lived elsewhere for much of the last seventeen years, reflections on my experience in the city have been enriched by that background. I like the words of the geographer Yi-Fu Tuan: "Long residence enables us to know a place intimately, yet its image may lack sharpness unless we can also see it from the outside and reflect upon our experience."[8] In analyzing notes from these sources, I documented a number of common themes, or aspects, of life in Las Vegas that provide a framework for the chapters to follow, and the stories of locals I met and learned about make up the narratives therein.

WHO, WHERE, AND WHEN

Prior to unveiling this portrait of Las Vegas, I need to address three concepts to stretch the canvas and apply the gesso, so to speak, to give a proper foundation on the "who, where, and when" of the work. Such a foundation lies in the answers to the following questions: What is a local? What is Las Vegas? Why do I choose the time frame I do?

I use the term "local" as a general reference to residents of Las Vegas. Such reference, when used as a noun, may seem somewhat pejorative, but it is standard within the Las Vegas vernacular. For example, casinos frequented by residents are commonly termed neighborhood or locals casinos. One of these, in fact—Sam's Town Hotel and Gambling Hall on Boulder Highway— used to advertise that it was "the place where locals bring their friends." Similarly, Strip casinos, golf courses, and amusement parks often advertise a "local rate." Residents also regularly use the word for themselves as a means of separation from the tourists, further underscoring the power of the insider/outsider dichotomy in Las Vegas.

Using "local" as a blanket title for all Las Vegans, however, is not entirely accurate. Some people, even longtime residents, do not deem themselves locals. Consider the following examples from conversations I had with four different people. Karl Marlin, originally from the Sacramento area, came to Las Vegas by way of Utah in 2000 with his wife and children. When I asked if he considered himself a local, he said no. He considers a local to be one who was raised here. Joanne Boyce, who came to Vegas from Baltimore more than ten years before Karl, and raised her children in the city, asserted that she, too, was not a local. She held instead an attachment to her former home, even after almost two decades. Ben Wychof, who came from Milwaukee in 2005, answered with a definite "yes" to my inquiry. He cited as evidence that his outsider friends ask him questions about his new place of residence. Then, somewhat jokingly (but extremely tellingly), he said, "I'll tell you where you really see the difference between a local and a tourist. It is in the [golfing] greens fees. Locals get a reduced rate. They make you show your driver's license and such. If you are from out of town, you pay through the nose."

Given such responses, it is difficult to determine a Las Vegas local precisely. From the first two responses, one might conclude that a local is either one born and raised in Las Vegas or who has spent many years living in the place. The third comment, however, disputes this definition. I tend toward a compromise position exemplified by a fourth conversation. Tracy Snow, a Kansas native who came to Las Vegas following her college graduation in 2004, answered my question as follows:

> Yes and no. When I meet new people here (especially those who are new to town), I find that the conversation inevitably turns into talk of restaurants or activities that are more "localized." . . . I think I like to show newcomers that there is more to the city than the Strip, kind of like a local's ambassador. And . . . I do have one local casino that I always go to. I have favorite grocery stores, restaurants, movie theaters, and things like that. When I first moved here, sometimes we would go down to the Strip just to walk around, for something to do. I don't do that anymore. The novelty wore off pretty quickly. In these ways I kinda feel like a local.
>
> But at the same time, I don't feel like one. A few weeks ago I was in Seattle for vacation. [When] asked where I was from, I would always reply, "Oh, I live in Las Vegas but I'm from Kansas." I don't know how long I have to live here before I will start answering that I'm from Las Vegas. I think if I were to move to a new city and was asked where I was from, I would still explain that I lived in Las Vegas for X amount of years, but I'm from Kansas. I like living here, but when I think about getting married and raising a family, I don't envision it happening here. Maybe if I did, I would feel more like I was from here.

Two important themes in Snow's response help to clarify the concept of the local: a time element and a sense of entrenchment in, or attachment to, the place. She has lived in Las Vegas long enough now that she has taken ownership of "local" things in the city and even acts as an ambassador to newcomers. At the same time, she is still a Kansan. But she might change allegiances if she were to raise a family in the city. Yi-Fu Tuan might summarize the dilemma of the Las Vegas local this way: "While it takes time to form an attachment to place, the quality and intensity of experience matters more than simple duration."[9]

What, then, is a local? The answer can come down to semantics, but I think the exercise can be illuminating and serves as a first lesson in a work on the personality of place. For Las Vegas the answer is as follows: Because this has been one of the country's fastest growing cities for decades, nearly everyone is from somewhere else. Normal, everyday conversations easily turn to the interlocutors' places of origin. On rare occasions in such discussions, one might meet a true native, but more than 96 percent of Las Vegans were born outside of Nevada, according to a 2007 survey.[10] Add to that Las Vegas's historically booming economy, plentiful jobs, and reputation for being able to "make it," and you have a recipe for a transient town. With Las Vegas's growth and transience, the experience of its residents is not so different from the broader contemporary American urban experience. We all move around more than we used to. And, because of such trends, many of us are "from" somewhere else. The concepts of "home" and "belonging" take on new meaning and importance in such circumstances.

In the end, when everything shakes out, two specific, small groups of people exist in Las Vegas—and in most cities—together with a larger, but less well defined middle group. This results in a continuum of belonging, what one local columnist called a "strata of locals." The "native" is born and raised in a place; the "newcomer," like Tracy Snow, is, early on in her residency, enthralled with the newness of place and hasn't quite found the "local" stasis; and, finally, the "local" is everyone else in the vast middle portion of the spectrum. The blurred edges of this middle group take in many natives and newcomers. Thus, "local" as it is used by Las Vegans—and by me in this book—typically refers to any person who resides within the Las Vegas Valley and who, therefore, experiences life within this unique place.

Defining the geographical space of Las Vegas is more straightforward. Like many large cities in the United States, Las Vegas is more than a single municipality. Most people agree that it consists of three incorporated cities— Las Vegas, North Las Vegas, and Henderson—plus the densely populated unincorporated portions of Clark County within the Las Vegas Valley. A few expansionists might include Boulder City within the metropolitan area. This community of sixteen thousand people lies to the southwest in Eldorado Valley on the other side of Railroad Pass, one of the handful of breaks in the Las Vegas Valley's surrounding mountains. Boulder City was built by the federal government to house workers at the Boulder (now Hoover) Dam project in the 1930s. Yet, its restrictions on gambling, regulated growth, unique history, and location out of the valley keep it separate—more of a bedroom suburb to the greater metropolitan area. I confined my interviews and my analysis to the Las Vegas Valley, and when I use the term "Las Vegas" or "the valley," it is in reference to this space. In speaking of the differences and similarities, conflict and cooperation among the different governing districts in the valley, I will refer to the municipality of Las Vegas as the "city of Las Vegas" in order to avoid confusion.

Multiple jurisdictions make it difficult to establish summaries from standard demographic, economic, or political data. For example, Las Vegas census data used to monitor growth is usually taken at the level of the Metropolitan Statistical Area (MSA), which includes the entirety of Clark County. Although the population in Mesquite, Boulder City, and the rest of Clark County outside the valley is relatively small, their numbers nonetheless impact the totals. To accurately portray my study area with numerical data throughout this work, I draw on various sources depending on the intended purpose. In addition to using US Census Bureau data, I also incorporate local data sets for Clark County including those available through the county's planning department, the University of Nevada, Las Vegas (UNLV), and periodic publications such as the *Las Vegas Perspective,* an annual with a decades-long record highlighting economic, business, and demographic information about the city.

Because my interest lies in present-day Las Vegas, this book largely confines itself to contemporary aspects. One reason for such a focus is to limit an otherwise huge task. Given the recent meteoric growth—with thousands of people moving to the city each month and new subdivisions, apartment buildings, and hotel-casinos constantly under construction—and the more recent lull in such activity, the city is always in flux. Some aspects of the place's personality are similarly changing. In fact, that change is an important element in the city's sense of place that I will discuss in a later chapter. My contemporary focus was also determined by my source material; the stories Las Vegans told me most often had to do with their current experience. Although longtime residents sometimes shared things about the past, such talk usually ended up bringing us back to what the place is like for them today and how it has changed. Because of the value and richness of such nostalgic perspectives and how they relate to life in the city today, I have included many of them in my analysis. Furthermore, because my own experi-

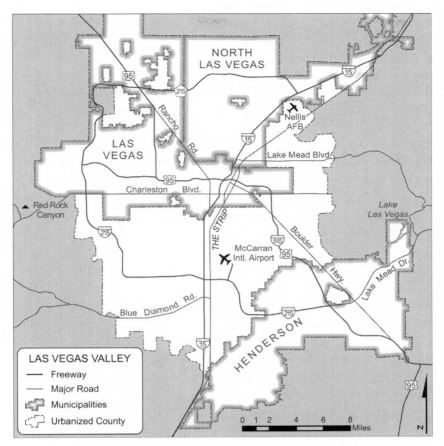

The Las Vegas Valley with major roads. All areas in the valley not within the municipal boundaries shown are considered unincorporated urbanized Clark County. Map by author.

ences in the city also play a role in this research, I have only the time frame of those experiences as reference: roughly from the late 1980s to the present.

Any writer wants to have their work remain timely and relevant. A contemporary analysis can quickly become outdated, of course, and in the case of a rapidly changing place like Las Vegas, the danger is amplified. Knowing this, I stress personality traits in Las Vegas that outlast shifts in growth rates or hotel construction cycles. Such traits might also serve as patterns for cultural and societal issues that could arise in other communities, regions, or nations. Aspects of contemporary growth that I examine, for example, likely will be outdated in twenty years, but the lessons of that struggle may warn and educate future leaders in Las Vegas and elsewhere as they confront related challenges. Historians and historical geographers provide lessons about today by analyzing events and people in a past place. So, too, can a "current history" of that place provide lessons for the future.

WHAT TO EXPECT

Describing something as complex and multifaceted as a city requires a framework. I have chosen to organize my rendering of the local's Las Vegas the way an artist would a painting. I will begin by sketching a layout in two chapters of context. Following this introduction, chapter 1 provides a snapshot of local history, and chapter 2 depicts the setting of Las Vegas, including its physical site and its geographic situation in Nevada and the American West.

The next three chapters contain several impressions, or broad-brush themes, that derive from, and are supported by, common points in my experience and remarks by local residents. Each chapter illustrates a major aspect of life in Las Vegas today. Chapter 3 will present a selection of stories about why people come to the Las Vegas Valley in order to illustrate the transient nature of the city. Chapters 4 and 5 emphasize the city's growth; I explore how residents cope and relate to the rapid changes that have taken place over the last two to three decades and then discuss traffic, a consequence of growth so prominent in local minds that it deserves its own chapter.

Building on the broad-brush impressions, the final three chapters create a series of narrower and more detailed "portraits" exploring how locals relate to the ever-powerful, ever-present "Vegas image." Chapter 6 is a presentation of how locals interact with the tourist corridor and how the city's twenty-four-hour nature affects the everyday lives of residents. Chapter 7 addresses the specific interaction between locals and the gambling and adult-entertainment industries, including the little-known (outside of the Las Vegas Valley) institution of the locals, or neighborhood, casinos. Then comes a chapter on religious life in a city known for its sin. Finally, a conclusion sums up the lessons learned through this look at place and explores the trajectory of a city at an environmental and economic crossroads.

Throughout the book, I have included a number of photographs. Some function as evidence to support a particular idea, while others are meant to illustrate an idea. To get to know a place, I feel that one actually must see it. In other words, visualizing a landscape that simultaneously represents and influences its people can contribute to our understanding of a sense of place. In the introduction to her influential book *The Death and Life of Great American Cities,* Jane Jacobs noted: "The scenes that illustrate this book are all about us. For illustrations, please look closely at real cities. While you are looking, you might as well also listen, linger and think about what you see."[11] The views I present within these pages are ones anybody might experience on a stroll through the city. And even though I provide a narrative of my observations and interpretations of a particular scene, my hope is that you, the reader, will view these images with an inquiring and analytical eye, doing as Jacobs exhorted and "linger and think about what you see" in this "real city."

MORE THAN JUST A PLACE

As an overall goal, I want to paint a picture of the insider's Las Vegas—what it means to live in a tourist city and what personality such a place holds for residents. My motives for doing so are manifold. Certainly, I hope that a reader who knows very little about Las Vegas, who may have visited the Strip once or twice or seen one of its many portrayals in TV or cinema, will come away from this book with a richer appreciation of life in Las Vegas, its unique and mundane aspects and the challenges and benefits facing its residents. At the same time, I hope that a Las Vegan who reads this might, after a certain passage, say, "You know, I didn't realize that about this town, but I see it happening," or "Wow! That's just how I might describe it." Such an outcome would mean that I did my job.

I also hope that this book is more than a simple portrait of a city. It is a portrait framed by and presented from a geographer's perspective: one where, to paraphrase the words of John K. Wright, the "encircling border [of terrae incognitae is] pushed back a little way" so that we might learn something about ourselves, that we might find, as did explorers of old, "a region not so greatly different from [our] own."[12]

Yes, at first glance Las Vegas has a character that is unique and, in some ways, extreme. And most Las Vegans, when asked, will articulate how distinct their life is from the tourist side of the city. At the same time, elements of the tourist image, landscape, and personality permeate nearly all aspects of life in the local's Las Vegas in fundamental and meaningful ways. So it is, I believe, in other places as well. The most obvious connections are to other tourist centers such as Orlando, New Orleans, Monte Carlo, and Paris. The argument can extend beyond this category, however, to a typical American city or even the most remote town. After all, an insider will always see a place in a different light than a visitor or passer-through. Herein lies my deeper hope for this book: the Southern Nevada metropolis can teach us much about other cities and towns, American culture, and culture in general. In the end, this book is about place: how people and cultures relate to place, how place shapes our lives, and how, in turn, our lives shape place. As we strive to understand *our* places, we can achieve a depth of awareness about ourselves.

One Hundred Years of Opportunity, Luck, and Rapid Change

H istory is easy to see in Las Vegas. Within an hour or so drive of down-town, one can hike on ancient volcanic formations, view sedimentary rocks that once formed the beds of Pleistocene lakes, photograph for-ests of bristlecone pines (one of the planet's oldest living things), see Native American petroglyphs, visit ghost towns and remnants of once-thriving min-ing communities, and observe ranchers practicing the centuries-old meth-ods of transhumance. In the city, older casinos—like the Dunes, the Sands, the Stardust, and the Desert Inn—are imploded to make way for the new. With the rapid suburban growth of recent years, locals are bound to see new intersections, freeway on-ramps, and subdivisions everywhere they go. Most Las Vegas residents, even the recent arrivals, have a personal story of, "I remember when that road was dirt and there were no homes past this inter-section, just desert." The connection to the past also extends farther back in time. In recent years, particularly with the celebration of the city's hundredth birthday in 2005, monuments, books, local radio shows, and community events help remind residents of earlier days in Las Vegas.

On a hot summer day in 2007, I experienced how visible the entire span of Las Vegas history can be. One of my first interviewees for this project, a mechanic by trade and rodeo cowboy by passion, reminded me that wild mustangs could still be seen up Red Rock Canyon and near Mount Charles-ton, favorite local playgrounds off to the city's west and north. So, to investi-gate and escape the heat in the valley, I took my family to Cold Creek, on the northern flank of Mount Charleston. Our path followed Highway 95, which rises to a point where one can view the entire city in the rearview mirror before then descending toward Indian Springs, the next in a seemingly end-less chain of valleys and ranges that makes up Nevada's topography. When I was younger the exit to State Highway 157 and the popular Kyle Canyon camping and hiking areas came some fifteen minutes after the edge of town. Today Kyle Canyon comes a mere two minutes from suburban development as the city has crept nearer its outdoor wilderness.

Another twenty-five miles on US 95 after the exit to Kyle Canyon, we turned onto the winding two-lane road that leads to the small mountain town of Cold Creek. We enjoyed the scenery of Indian Springs Valley behind, Mount Charleston and the Spring Mountains ahead, and, on a short hike, saw signs that horses had traversed the area recently. Then, heading back down the mountain, we spotted the mustangs and spent the next hour watching them drink from an irrigation ditch and graze on adjacent forage. Even though today much of the mustang population is controlled by the Bureau of Land Management, the sight drew my mind away from the bustle of the city, giving me a glimpse of a time when these majestic, wild creatures were more abundant and the first locals were making this desert oasis their home. Then, on our return home, that glimpse turned into a vision. Around five miles north of the Vegas Valley's edge and about two miles east of Highway 95 sits the obscure village of Corn Creek. Most people would never notice the town, but to me it appeared just as Las Vegas must have been in 1905, a small hamlet of little consequence to the outside world, established near a water source in the center of a desert valley.

On my trip up Highway 95, I caught a glimpse of Southern Nevada's natural history through the magnificence of the wild horses, the first page of the modern city in my vision at Corn Creek, and a more recent page in the sprawling suburban development. My goal in this chapter is to fill in the gaps. As background for the rest of the book, I will lay out a "shotgun history," presenting key events in one quick burst.

BEFORE LAS VEGAS

The valley's earliest residents were probably Paleo-Indian hunters more than eleven thousand years ago, when the brown Mojave Desert of today was a lush, well-watered landscape. Archaeologists claim that these ancient peoples used the area surrounding Tule Springs, the site of present-day Floyd Lamb Park a few miles northwest of downtown Las Vegas, as a base. Following their demise amid climatic changes between 8,000 and 5,500 B.C., nearly two millennia passed before the next group moved into the valley, a group known for their use of caves in the surrounding mountains. Next, around 300 B.C., a people called "Basket Makers" occupied the southern Nevada desert. They lived in pit houses near natural springs; gathered nuts, seeds, and other desert plants; and hunted rabbits, birds, desert tortoise, and the occasional deer or bighorn sheep. Petroglyphs in the sandstone rocks of Red Rock Canyon and Valley of Fire State Park document some of these activities.

Generally sedentary, the Basket Makers migrated only when the areas around their dwellings failed to provide the necessary sustenance. Some scholars have wondered if these people merged with Puebloan groups from the Four Corners region, turning for evidence to locally discovered adobe structures and clay dishes dated to around A.D. 500 similar to artifacts from the latter cultural group.

After about A.D. 700 the Southern Paiutes were the only inhabitants of the Las Vegas Valley until the Europeans came. Maintaining an existence similar to the Basket Makers, the Paiutes principally hunted and gathered local species, but also farmed to some extent. They resided near springs during the winter months, where they constructed elliptically shaped wickiup shelters from brush and branches. In the summer they retreated into the mountains surrounding the valley to escape the heat. Many southern Nevada Paiutes still live in Las Vegas on land donated to the tribe by Helen Stewart, a later rancher in the valley, and in reservations northeast and northwest of town.

GETTING A NAME

The advent of Las Vegas as a place on New World maps was a direct result of the Old Spanish Trail. Explorers such as Francisco Garces and Francisco Escalente in the 1700s and Jedediah Smith in the early 1800s helped to establish this trade route from Santa Fe to Southern California settlements, but they avoided the direct path through the arid lands of what is now southern Nevada in preference for a road that kept travelers close to the Colorado River. The sixty-mile stretch of desert between the Muddy River and Las Vegas was so inhospitable, in fact, that it earned the name on later maps of *jornada de muerto,* or "journey of the dead man."

In the years following Smith's explorations, a series of events facilitated the discovery of Las Vegas and the forging of a new version of the Spanish Trail. On November 7, 1829, Antonio Armijo, who headed the first caravan to transport commercial goods from New Mexico to California, discovered springs several miles from the lifeline of the mighty Colorado or its tributaries. Armijo had camped on Christmas Day near the present-day site of Mesquite, Nevada, and sent scouts in search of watering places to the southwest near the Mojave River. But, the jornada de muerto and other unseen arid landscapes lay between the caravan and its eventual goal of the Mojave. While the scouts were away, Armijo's party continued south along this older trail until they reached the mouth of the Las Vegas Wash and its confluence with the Colorado. Here Rafael Rivera, one of Armijo's scouts, rejoined the party and reported finding springs at the wash's head. Although Armijo's group did not actually visit the springs on their trek through the valley, knowledge of such water in the area attracted others. One of the groups who subsequently made the direct trip through the area in the early 1830s named the oasis "Las Vegas," commonly translated as "the meadows," but literally meaning "the fertile valley." Either translation fits.

The old and new versions of the Spanish Trail vied for popularity in the late 1830s and early 1840s. Then, in 1844, the legendary John C. Frémont changed this status when his famous topographic expedition passed through the southern Nevada desert. The party camped at Big Springs—the present location of the Las Vegas Springs Preserve at the intersection of US 95 and Valley View Boulevard and across the street from the aptly named Meadows Mall—and subsequently placed the Las Vegas name on their official map of the journey. They were the first to do so. Frémont also named the tall peak in the Spring Mountains to the west "Charleston," to honor his hometown in South Carolina.

Although Frémont's map made the northern variant of the Old Spanish Trail through Nevada more popular, the stretch through southern Nevada remained threatening. S. N. Carvalho, one of Frémont's artists on a later expedition, wrote the following of the passage between the Muddy River and Las Vegas: "It was not difficult to follow the trail; in one hour I counted the putrid carcasses of nineteen oxen, cows, mules, and horses; what a lesson to those who travel over such a country, unadvised and unprepared."[1] Still, as more people traveled west, this western arm of the old trail, roughly following present-day Interstate 15 from Parowan, Utah, through Las Vegas to Los Angeles, found new life as a corridor for mail, freight, and immigrant groups traveling between Salt Lake City and Los Angeles. In fact, early traffic between settlements of the Church of Jesus Christ of Latter-day Saints in Southern California and that organization's headquarters in Salt Lake City along this "Mormon Trail" was sufficient to attract the first non-Indian settlers to "the meadows."

MORMONS AND RANCHERS

Many Las Vegas residents today will tell you that the city was settled by the Latter-day Saints. This claim often comes as a point of evidence in their argument that Sin City is a good, wholesome place. Over 100,000 Mormons live in the Las Vegas Valley today, and the city, as a whole, has a thriving religious culture, but the claim that the Mormons settled the place is only partially true.[2] A group of them in 1855 *were* the first non-Native people to set down roots, but those roots were only semipermanent. The majority of this party abandoned the valley a year and a half after they arrived, leaving the area to a handful of non-Mormon ranchers for another fifty years.

Under the leadership of Brigham Young, the church had expanded its influence throughout the West since arriving in the Salt Lake Valley in the 1840s. Young established colonies throughout the Utah and New Mexico Territories, and, with the goal of maintaining a corridor between Salt Lake City and the Pacific Ocean, the Mormons settled in San Bernardino, California. In 1855, a group of around thirty missionaries established an outpost along the Mormon Trail, near the life-giving springs of Las Vegas. At that time this place fell in New Mexico Territory just south of the Utah Territory line as it extended west to California. (Nevada would not become a state until 1861.)

The colonists, under the leadership of William Bringhurst, arrived in Las Vegas on July 16, 1855. Not wanting to displace the Paiutes living in that area, they built camp a short distance from Big Springs. This site lay on a low shelf overlooking the eastern part of the valley and gave them access to water via Las Vegas Creek, a stream flowing out of the springs. The settlers constructed homes, farmed, and built a fort for protection. They also established a post office called Bringhurst, New Mexico Territory, since another Las Vegas in that territory had already taken the name. Remnants and reconstructions of the fort can be seen today at a state park at the intersection of Las Vegas Boulevard and Washington Avenue near Cashman Field.

Social problems caused the mission's demise. One was a leadership clash after Brigham Young sent Nathaniel Jones to Las Vegas in 1856 to head a lead-mining operation in the nearby mountains. Jones and Bringhurst argued over their respective roles, eventually leading to the original leader's dismissal. The ore was also low in quality and difficult to transport. These issues, plus struggles in agricultural production, lack of success in teaching the Paiutes, and loss of crops to Indian thieves, led Young to close the mission in February 1857, only nineteen months after its establishment.

Enough travelers continued to stop at the springs on their journeys to and from California that Albert Knapp, one of the original members of the Las Vegas mission, returned in 1860 or 1861 and set up a store in one of the fort buildings. Later, Albert's brother William took over this operation and apparently remained for some time, as evidenced by his dealings with later ranchers who would inhabit the valley.

Soon after Knapp's return, Las Vegas had a short-lived name change and, with it, a place in Civil War history. With a threat of Confederate invasion of California from Texas across New Mexico Territory, newspapers pushed for Union soldiers to be stationed at Las Vegas. Union general James Carleton liked the concept, but opted for deception instead of actual troops. During the winter of 1862 Carleton publicly announced the Union's intentions to reoccupy the Las Vegas fort and rename it Fort Baker. He never did so, but used the propaganda to cover his own push across the Southwest and into Texas. As Ralph Roske has written, "Carleton's ruse worked better than he had planned. Many historians, misled by [his] newspaper plants, have identified Fort Baker in their books as a real installation."[3]

In the final decades of the nineteenth century Las Vegas became home to a handful of ranching families. A friend of William Knapp, Octavius Decatur Gass, was the first. He came west in 1850 in search of gold and, not finding success, took over the Mormon fort and its overgrown fields in 1864. Gass was quite successful in this new career. In 1872 he controlled 120 acres surrounding the old fort, but by 1878 his holdings had grown to 640 acres and virtually all of the valley's water. He ran fifteen hundred cattle and irrigated crops as diverse as wheat, oats, barley, beans, grapes, apples, and peaches. The Las Vegas Ranch, as Gass's holdings were known, served markets in Arizona and Colorado and catered to the many travelers passing through the valley. He also encouraged an old Ohio friend, Conrad Kiel, to take up the Mormons' farm north of the fort in present-day North Las Vegas. The canyon leading to Mount Charleston mentioned earlier acquired the Kiel name because of a lumber mill the family operated there. The canyon's name was eventually changed to the more common "Kyle."

When New Mexico Territory was subdivided in 1863, Las Vegas lay within the giant Mohave County that occupied the northwest corner of the new Arizona Territory. The next year, O. D. Gass became an Arizona legislator, eventually being elected that body's presiding officer. Gass lobbied successfully for the creation of a new Pah-Ute County out of Mohave County, which took in his interests in the Las Vegas Valley. Gass was less than pleased in 1867 when Nevada accepted lands west of the Colorado River in Pah-Ute and Mohave Counties as part of that state's new Lincoln County. In fact, most Pah-Ute County residents in the Las Vegas Valley and nearby Mormon settlements in the Moapa Valley near the Muddy River still considered themselves part of Arizona until 1871 when Arizona legislators finally relinquished claims to this area.

With the boundary change, Gass lost political power, which added to a suite of financial difficulties for the rancher. His military customers stationed at nearby Callville on the Colorado River claimed that his prices were too high, he was accused of stealing horses, he owed back taxes to his new state, and he battled litigation on rival mining claims in California. Gass mortgaged his ranch twice between 1876 and 1879 to cover debts. He paid back the first loan, but was unable to cover the second. So, in May 1881, Gass forfeited his large holdings in the valley to a Pioche, Nevada, businessman named Archibald Stewart, and left the ranch a month later.

Taking over operations on Las Vegas Ranch, Stewart, his wife Helen, and their four children became the valley's newest permanent residents. Selling commodities in area mining markets, Stewart prospered for several years before his violent and mysterious murder on July 13, 1884. On that day, Stewart had left home, rifle in hand, to confront Schuyler Henry, a neighbor who was spreading gossip about Helen. Henry claimed to have killed Stewart during a fair gunfight at the Kiel Ranch. Helen, however, believed that the death was part of a conspiracy instigated by nearby rancher Conrad Kiel as a reprisal for Stewart's foreclosure on the ranch held by Gass, a friend of Kiel. A jury dismissed charges against Kiel and Henry for lack of evidence, but sixteen years later a new element reignited the murder mystery. Helen Stewart's son, Will, accompanied the Stewart ranch foreman to purchase tobacco at the Kiel Ranch, which had continued operations under Conrad's two sons, Edwin and William, after their father's death in 1894. They claimed they found both owners dead when they arrived. A jury determined the deaths a murder-suicide, but a later archaeological team found evidence in the victims' bones that suggested a double murder, possibly in revenge for Stewart's death. All three crimes remain unsolved.

Helen Stewart took over ranch operations, which she continued for the next twenty years. After trying to sell on several occasions, she finally did so in October 1902. A wealthy senator and mining magnate from Montana purchased the property for $55,000 with plans to build a railroad through the site. Helen Stewart is remembered fondly in local history. Known as "the First Lady of Las Vegas," she helped start the Mesquite Club, a women's organization still active today, and has been recognized for her collection of Paiute Indian baskets. Both she and Gass have downtown streets named for them. The life they enjoyed in the Las Vegas Valley, however, would soon change.

A NEW WESTERN TOWN

It was a hot day, May 15, 1905, when twelve hundred lots went on the auction block in the newly platted railroad town of Las Vegas. William Andrews Clark, the Montana senator who had purchased the Stewart Ranch, had a grand scheme in mind. At a time when rails were approaching their maximum density in the United States, no trains yet passed between Salt Lake City and Los Angeles directly. In competition with the Union Pacific company (their interests would eventually align in a compromise) Senator Clark pushed his San Pedro, Los Angeles, and Salt Lake Railroad along the same general path as the Mormon Trail. The line was completed January 30, 1905. Needing a division point to provide repairs to trains and a respite to passengers, Clark selected Las Vegas because of its centrality, water supply, and availability of lumber in nearby mountains. A frontier tent city began to rise from the desert floor.

Clark's original townsite took in the blocks of present-day downtown Las Vegas bounded by Stewart Street on the north, Garces Street on the south, Main on the west, and Fifth Street on the east. Fremont Street served as the hub of business and commerce.

Clark was not the first to try. John T. McWilliams, a civil engineer working in southern Nevada since the 1890s, also saw potential for a town. He had purchased an eighty-acre plot from Stewart and began his own Las Vegas townsite just prior to the 1905 auction. A shanty ragtown on the north side of the SPLA&SL Railroad, McWilliams's townsite drew some interest, but his lack of water rights and sufficient capital soon spelled doom. Many northside inhabitants moved their tents and shacks across the railroad tracks to the Clark townsite following its establishment. A fire late in September 1905 destroyed what was left of McWilliams's settlement, located in today's West Las Vegas neighborhood.

From its beginnings, Las Vegas depended on outside forces for its survival. Initially, of course, that dependence revolved around the railroad. Many Las Vegas residents were beholden to jobs the railroad provided, while others prospered because of marketing opportunities it offered. The merchant trade area spread to several nearby mining and agricultural settlements, such as Beatty, Goodsprings, and the Moapa Valley. The railroad also provided, through its subsidiary the Las Vegas Land and Water Company, municipal water and other infrastructure.

As the local population increased from 945 residents at the 1910 census to 2,304 in 1920, Las Vegas enjoyed relative stability. In 1909 the state had made it the seat of Clark County, which was carved out of Lincoln County, and incorporated Las Vegas as a city in 1910. This prosperity ended in 1922, however, when Las Vegas rail workers walked off the job as part of a nation-wide strike. Having purchased Clark's interests in the SPLA&SL Railroad, Union Pacific leaders were not happy with the walkout. They cut off electricity to the town, brought in strikebreakers to continue the work, and eventually, when the strike ended, moved the local repair shops north to Caliente. Company officials claimed these actions were not punitive, and rail traffic continued to move through the town, but Las Vegans nonetheless felt the effects. After all, the railroad had employed 15 percent of the townspeople and affected all the rest. As a result, the celebration of the Roaring Twenties heard elsewhere in America was largely silenced in southern Nevada.

During the slump of the 1920s Las Vegas tried to take advantage of another railroad-related possibility: visitors and newcomers. As early as 1911 the Las Vegas Promotion Society had been advertising the town to outsiders for its agricultural potential and free-flowing artesian water. Boosters also promoted tourism by touting the area's favorable climate. Rail passengers might spend their layover at one of the several hotels on Fremont Street or refresh themselves in one of the bars on Block 16, an original railroad auction parcel set aside as the only place one could purchase liquor. Some of the visitors patronized brothels attached to Block 16 establishments, which would provide services for many years to come. This red-light district was disliked by many residents, of course, but became part of the fabric of the budding community. Hal Curtis, an early resident, recalled how he and other boys made deliveries there. He remembered, "them whores was good people."[4]

Other tourism-centered elements added color to the town's frontier character. Gambling was legal and prevalent throughout the West in the 1800s. Nevada passed some prohibitionary legislation in 1910, but these laws were largely ignored and casino games abounded locally throughout the 1910s and 1920s. By 1931, amid declining revenues in mining, Nevada legislators passed laws to legalize wide-open gaming. In addition, the state relaxed divorce laws, which Las Vegas used for its benefit. Las Vegas also is remembered for its blatant disregard for liquor laws. Its small size and isolated location made it relatively easy for bootleggers to practice their trade away from the watchful eyes of federal agents.

WATER AND WAR SAVE LAS VEGAS

When the Boulder Canyon Act was passed on December 21, 1928, authorizing the damming of the Colorado River just thirty miles from Las Vegas, the local atmosphere was jubilant. It was a chance to add many new jobs. One resident recalled, "bootleg liquor just flowed like water."[5] Another added, "There was people that got lit that never had taken a drink before."[6] The following day more than two hundred people made a prayerful pilgrimage to the dam site and "gave thanks for the blessings vouchsafed to them and to the community."[7] The coming of Boulder Dam, later to be renamed after President Herbert Hoover, gave plentiful reason for rejoicing. The largest public works program in US history (apart from the Panama Canal) had just been placed in their lap.

Las Vegans were wildly optimistic at the benefits they might reap from the project. With a population of only 5,165 reported from the 1930 census, townspeople anticipated an explosion to between 25,000 and 100,000 people during and directly following dam construction. They also expected that cheap power from the dam would bring an infusion of millions of dollars in economic development, jump-starting them as one of the great industrial centers of the West. Moreover, the city hoped to host the thousands of construction workers to be employed at the dam. Soon, all of these dreams had to be tempered. First, citing Las Vegas's disregard for Prohibition and its wild frontier character, federal officials insisted on another, more wholesome housing option for dam workers. The result was Boulder City, a government-owned new town eight miles from the dam site and twenty miles from Las Vegas.

The dam did, however, mark a turning point in the fate of Las Vegas. Its construction work enabled the city to sail through the tumultuous storm of the Great Depression with few damaging effects. Through the early years of the Depression, in fact, Las Vegas experienced a real estate boom, with businesses reporting increases over previous years, new neighborhoods springing up away from the town center, and several infrastructure improvements as the city gained new status as "Gateway to the Boulder Dam." Boulder City contributed to the stabilizing economy as hundreds of its residents made the short drive to Las Vegas each day for everything from buying milk to seeing a movie at the El Portal to patronizing the whorehouses on Block 16. Hundreds of thousands of tourists from outside the region also passed through Las Vegas in the 1930s on their way to see Boulder Dam under construction, three hundred thousand in 1934 alone. Indeed, Las Vegas had become almost wholly dependent on the dam for its survival. Al Cahlan, then editor of the *Las Vegas Evening Review-Journal,* put it into perspective. Without the dam, he wrote, "Las Vegas would be in a Hell of a fix."[8] At the same time, the gateway city soon faced struggles of a different sort as it was forced to deal with thousands of jobless men and families, poor and hopeless in the depressed economy, who had come here in search of work. Although the number of immigrants is unknown, many thousands more than the dam could employ came to the valley, bringing with them crime, hunger, and sickness to which city residents were forced to respond.

Despite the economic prosperity brought by dam construction and an increase in tourist trade, Las Vegans' grandiose plans were largely unfulfilled when Boulder Dam was dedicated in 1935 by President Franklin D. Roosevelt. Just as residents had felt how much they depended on the railroad when the city lost the repair shops in 1922, they now felt a similar slump following completion of the dam. New Deal money continued to provide jobs and support local infrastructure improvements, and tourists continued to come to see the newly completed dam, but gradually all this business declined. To compensate, townspeople applied a lesson learned during the construction era: they could sell their city as a tourist destination. Proactive efforts began quickly, but surprisingly this boosterism did not focus on gambling. Instead, their signs and brochures proclaimed: "Las Vegas Nevada. Still a Frontier Town." Specifically, city leaders took advantage of Lake Mead (the reservoir created behind Boulder Dam) and hosted boating events starting in 1935. That same year the Las Vegas Elks Lodge began Helldorado, a parade and rodeo celebration with a frontier and mining theme. The region still struggled, however, as the Depression dragged on, until the US entry into World War II brought major federal investment that saved the city once again from suffering the ghost-town fate of so many Nevada communities.

The most notable impacts of the war were found on the northern and southern ends of the Las Vegas Valley. Under the encouraging arm of Nevada's senior senator, Pat McCarran, the city's small airport northeast of town was offered to the US Army Air Corps for development of the Las Vegas Army Air Field. Eventually becoming what is today Nellis Air Force Base, the airfield served as a gunnery school for over fifty-five thousand aviators during the four years of wartime involvement in Europe and the Pacific. The new airbase brought the incorporation of Las Vegas's first sister city in the valley (Boulder City is in Eldorado Valley). North Las Vegas had gotten its start in 1917 when Tom Williams purchased a 140-acre tract a little more than a mile north of Fremont Street for eight dollars an acre. He provided crude infrastructure, graded roads, and subdivided most of this land to sell to interested settlers. Although his intentions were to create a free and lawful community (he considered Las Vegas too lawless), more than a third of the eighty lots were purchased and put to use by moonshiners. Then, during the early 1930s, North Las Vegas became known for its "Hoover City" of tents and makeshift settlements for unemployed workers hoping for jobs at the dam project down the road. With the big influx of military personnel during the war, however, it was only a matter of time before North Las Vegas incorporated in 1946.

On the south end of the valley, a second wartime legacy created another new city in the valley. Even before American intervention in the war, a need for magnesium for use in bombs and aircraft led US military planners to southern Nevada. In 1939 they found large ore deposits in the local mountains. Senator McCarran took over from there. Pointing out Las Vegas's easy access to Lake Mead water for cooling purposes and Hoover Dam electricity to power the operation, he and his allies successfully sited a magnesium plant at the base of Black Mountain, roughly ten miles southeast of the city. Financed by the federal government, Basic Magnesium, Inc. (BMI) brought in between 3,600 and 6,500 workers from all over the country, including a sizable black population from Fordyce, Arkansas, and Tallulah, Louisiana. A place called "Basic Townsite" was created to house the workers, its name eventually changing to Henderson after a former senator from Nevada and board member for the Reconstruction Finance Corporation, Charles Belknap Henderson.

The airfield and BMI brought new life to Las Vegas both during and after the war. Workers from both places, as well as soldiers from other wartime bases as far away as San Diego, flocked to the city to enjoy the relaxed atmosphere of drinking and gambling. The overly relaxed morals on Block 16, however, were too much for army leaders. Not wanting to lose the base or the traffic it brought to gambling houses on Fremont Street, city commissioners conceded to a military ultimatum and officially banned prostitution. They also agreed to strict operating hours for casinos, bars, and liquor stores, each closing at midnight on weekdays and 2:00 A.M. on weekends. Las Vegans began to realize that carefully controlled gambling-based tourism could be an important part of their economic future.

Las Vegans also saw their city's residential scene change and expand as a result of wartime funding. The valley's population was 10,389 at the 1940 census. By 1950 new industrial and military developments had brought this count to 24,624 people in Las Vegas proper and 33,918 within the valley.[9] Housing needs for this influx led to new additions east and south of downtown, including the neighborhoods of Mayfair, Biltmore, and Huntridge. Into the 1950s and 1960s, most of this residential growth followed the extension of the tourist scene out of downtown and south along a lonely strip of US 91.

As the population boomed, entrepreneurs saw potential for entertainment clubs outside Fremont Street in downtown Las Vegas. The Pair-O-Dice and the Meadows Club were early attempts at such expansion in the 1930s. Located just south of the Las Vegas line on Highway 91 (an extension of downtown's Fifth Street that would be renamed Las Vegas Boulevard) in an unincorporated town called Paradise, the Pair-O-Dice saw success as a small gambling and dancing club. A favorite of locals and tourists, the Pair-O-Dice (later renamed the 91 Club) was nothing in comparison to the modern resorts that would join it a decade later. Tony Cornero, a gambler from California, took advantage of gaming's legalization in 1931 to open up the other major non-downtown establishment. His Meadows Club opened on Boulder Highway (the eastern extension of Fremont Street) that same year, but a fire closed it down just a few months later. As an outsider who came to

View of Fremont Street's Glitter Gulch looking east, around 1950. Photo courtesy of UNLV Libraries, Special Collections.

Las Vegas to do legally what he had been doing elsewhere illegally, Cornero would serve as a model for future builders of the casino landscape in the valley. Although another of those outsiders, Bugsy Siegel, often gets the credit for the rise of the Las Vegas "Strip," credit in fact should go to Thomas Hull, who in 1941 built the first successful resort hotel and casino on the corner of Highway 91 and then San Francisco Avenue (Sahara Avenue today).

A GROWING TOURIST CITY

Hull came to Las Vegas because local boosters convinced him that an expansion of his California chain of El Rancho hotels would be profitable for both parties, but nobody at the time realized the impact El Rancho Vegas would have on the city's future. Instead of selecting land near the downtown casino core, Hull purchased a large tract of desert outside of the city limits. Hull could thereby be free of burdensome city taxes and regulations. More important, Hull had enough space here to build a series of cottages and motel-style rooms, a huge casino, an expansive parking lot, and other resort amenities. In this way the El Rancho in April 1941 created a new model for casino resorts on Las Vegas Boulevard. Taking Hull's basic idea but throwing out the western theme pervasive in Vegas resorts of the day, Benjamin "Bugsy" Siegel soon set out to build a lavish "carpet joint" named for the colorful flooring seen in Miami resorts. He aimed his Flamingo resort at wealthy California and Florida patrons, and although it struggled for several years after opening in 1946, the hotel and resort eventually went on to financial success under Gus Greenbaum. Hull had set the mold, but Siegel's Flamingo became the flashy, frilly symbol of what Las Vegas was becoming. By the end of the decade, other neighbors on Highway 91 included the Last Frontier (1942), Club Bingo (1947), and the Thunderbird (1948). Guy McAfee, who ran the Golden Nugget downtown and the 91 Club, is credited with naming Las Vegas Boulevard "the Strip." In its 1940s form, the resort corridor reminded him of Sunset Strip in Los Angeles where he once had been a police officer.

With Strip development in full force by the late 1940s and continuing success at Glitter Gulch establishments on Fremont Street, the value of casino tourist gambling became undeniable. So, Las Vegas businesses rallied around a Chamber of Commerce effort called the Live Wire Fund, which provided a "war chest" for marketing purposes. In the postwar years, the chamber hired a series of national advertising agencies to give their city a national presence. Las Vegas also received free publicity from "drop-in," or "parachute," journalists who found stories about this city that embraced vice outwardly shunned

TOP: El Rancho Vegas. Opened in 1941, it was the first casino-resort on the Las Vegas Strip. Photo courtesy of UNLV Libraries, Special Collections/UNLV Center for Gaming Research.

BOTTOM: The Flamingo, Bugsy Siegel's flashy "carpet joint," which opened in 1946. Photo courtesy of UNLV Libraries, Special Collections.

in the rest of the country. Atomic testing at the Nevada Proving Grounds, northwest of the valley, began in January 1951, and made the city even more recognizable. Indeed, resorts hosted parties preceding blasts, which could be watched from atop the city's highest buildings. All this postwar marketing and publicity proved successful. By 1952 gambling had become Nevada's top industry, beating out mining and farming.

Las Vegas promoters sought to supplement the tourist and gambling visitor numbers with targeted marketing to conventioneers, honeymooners, and people seeking divorces. Smaller visitor counts between Mondays and Thursdays were a particular concern, and a convention industry was seen as a potential solution. Construction on the 7,500-seat Las Vegas Convention Center with 90,000 square feet of exhibition space was completed in 1959, a major supplement to the hotel meeting spaces that had to suffice in earlier years. The convention trade is stronger than ever today, coordinated and promoted by the Las Vegas Convention and Visitor's Authority (LVCVA). Promoters also had success attracting honeymooning couples. Nevada's liberal laws made marriage a simple proposition, and the city had hosted more than twenty-nine thousand weddings annually by the end of the 1950s. Family and friends accompanying the happy couple would also spend money, further boosting revenues for the city.

City boosters knew from past experience that the number of visitors alone was a narrow base to fuel their growing city, and so they worked hard to maintain local military and industrial resources. As war ended, leaders relied on the influence of their longtime US senator Pat McCarran to save the airfield. Success came in 1950 when the site became a permanent base under the auspices of the newly formed US Air Force and was renamed for William H. Nellis, a native of southern Nevada shot down over Belgium in 1944. Nellis Air Force Base originally sent fighter pilots to the combat zones of Korea. Today, it is a hub for fighter-pilot training and continues to play an important role in the economy and personality of the city. The city logo for North Las Vegas, in fact, shows an image of the B-2 stealth bomber.

Las Vegas also looked to nontourism benefits brought by atomic testing at the Nevada Proving Grounds, more commonly known as the Nevada Test Site. As the government played down any ill effects from aboveground testing there, Las Vegans enjoyed the added jobs. Testing went below ground after 1962, where it remained until all such activity was discontinued in the 1990s. Still in operation today, the Test Site was given a new name in August 2010, the Nevada National Security Site, reflecting a new primary mission to maintain the country's stockpile of nuclear weapons, and secondary roles to manage and train for potential nuclear response and test technologies intended to support nonproliferation.

The third arm of economic diversification in the valley came with continued activity at Basic Magnesium's industrial complex. Like the airfield, BMI was nearly lost at war's end. But through the fortitude of local boosters, support from both state and federal governments, and private companies interested in possibilities afforded by the established industrial site, BMI (today the Black Mountain Industrial Complex) continued operation and is still a noticeable feature in the southeast portion of the valley.

Postwar growth in tourism, government projects, and industry brought explosive expansion of the city's population. Census counts in the valley more than tripled during the 1940s and nearly did so again through the next decade when greater Las Vegas became home to more than 95,000 residents by 1960. With this growth came inevitable problems. During the 1950s, such struggles involved the provision of adequate water, telephone and electricity service, and streets and schools for new neighborhoods. The city's expansion rate simply outpaced the ability to supply necessities.

One of the legacies of the Strip from a municipal perspective was how it directed valley growth along its flanks, south and east of downtown. As the early Strip resorts were joined in the 1950s by the Desert Inn (1950), Sands (1952), Sahara (1952), Dunes (1955), Riviera (1955), Hacienda (1956), Tropicana (1957), and Stardust (1958), suburbs began to fill the county land to the east of Las Vegas Boulevard. Maryland Parkway, a continuation of what was 12th Street downtown running parallel to the Strip, became the focal point of that growth, hosting new shopping centers and office buildings as such activity moved away from its traditional home on Fremont Street. On the southern end of the parkway, at its intersection with Tropicana, the first buildings of Nevada Southern University were constructed in 1957, a campus that would become the University of Nevada, Las Vegas, twelve years later. Las Vegas envied such growth south of its city limits—although good in many ways, it brought no tax money to Las Vegas proper—and officials tried many times to annex the area. Resort owners would not have it, however, and the Strip and most land in the valley south of Sahara Avenue (excepting Henderson) still remain part of unincorporated Clark County.

One of Las Vegas's many nicknames is "the town the gangsters built." Although this tag is not completely accurate, much of postwar Las Vegas was connected with organized crime in one way or another. Early mob influence began in the 1930s when legalized gaming returned to the Nevada scene. More notable, however, was Bugsy Siegel's deep connections to the eastern mafia; his financing of the Flamingo was done with mob money. Siegel's foray initiated an era of intense cash flow from mob and mob-connected sources into the local economy that created much of the tourist landscape between the 1950s and 1970s. Success at the Flamingo, Dunes, Stardust, Riviera, Sands, and Tropicana can be linked directly to characters such as Gus Greenbaum, Moe Sedway, Meyer Lansky, and Moe Dalitz, all of whom were "connected."

Off the Strip, the Boulevard Mall (the valley's first), Sunrise Hospital, and surrounding office buildings were all projects executed by Paradise Development, which was headed by Moe Dalitz along with Merv Adelson and Irwin Molasky. Financing for these projects and others along the now bustling suburban stretch of Maryland Parkway came in the 1960s through the Teamsters' pension fund and Jimmy Hoffa, a former associate of Dalitz. In fact, Hoffa's fund was a major source of capital for the Bank of Las Vegas (later Valley Bank and eventually part of Bank of America), which was the only source of "legitimate" financing for casinos until the late 1970s. With so much positive impact on the economy, both in suburban projects and jobs provided by the resorts on the Strip and downtown, it is no wonder that the presence of the mob was tolerated, and even accepted, within the community.

Mob influence waned in the late 1960s, although some evidence of gangster activity remained into the early 1980s. Skimming operations and other illegal activities in the Aladdin and Tropicana casinos made headlines in the 1970s. Frank "Lefty" Rosenthal, a Chicago mobster, ran a large skimming operation at the Stardust. He, along with Anthony "the Ant" Spilatro, a violent enforcer for Rosenthal's gang, were main characters in the 1995 Martin Scorsese film *Casino*. The legacy of the mob is certainly recognizable today: it is celebrated in a museum occupying the former post office and courthouse building downtown, and Mayor Oscar Goodman, a defense attorney who made his career representing mobsters (playing himself, in fact, in *Casino*), finished his third term in 2011 as one of the most popular mayors in city history. Still, blatant mob involvement in building and operating casinos largely ended in 1969 when Nevada legislators passed a law to allow corporate ownership of casinos. Previously, such control was impossibly difficult because of a restriction that each shareholder must hold a gaming license in the state.

CHALLENGES AND BENEFITS OF GROWTH

The postwar boom in Las Vegas continued through the opening of a new century. During the 1960s, neighborhood expansion made its way farther west of the railroad toward major arteries of Valley View and Decatur Boulevard, the Test Site acting as a magnet in that direction for employees wanting to shorten their commute. Residential growth in the 1970s, however, continued in a southward direction along the Strip, east of the railroad tracks between Las Vegas Boulevard and Boulder Highway. Double-digit inflation in the early 1980s and tragic events such as the MGM Grand fire that killed more than eighty people in 1980 brought temporary lulls. A similar drop in revenue occurred following the attacks of September 11, 2001. But overall, the Las Vegas economy has remained resilient, and the community did nothing but grow and grow some more. The Census Bureau reported the Clark County population at 273,288 in 1970 and 463,087 in 1980. In 1996 the county hit the one million mark. And, despite economic hardship and tapering of the city's meteoric expansion, the 2010 census found Southern Nevada's population just shy of two million.

The explosive growth of the last six decades, however, brought its share of problems. A quick listing of the challenges that Las Vegas residents have faced in recent decades along with the benefits they've enjoyed provides a necessary context to the city's personality in the last quarter century that is the focus of subsequent chapters. Annexation is one major issue accompanying the city's expansion. City officials long have desired to annex more of the valley south of their borders, but the county continues to block such moves. Consolidation of the cities and county into a single governing body is another option, also championed by city of Las Vegas leaders. Clark County, Henderson, and North Las Vegas governments rejected this idea early on, too, although in 1973 the county sheriff and the Las Vegas city police department consolidated into the Metropolitan Police Department, commonly called Metro. The police for Henderson and North Las Vegas, however, retain separate jurisdictions. Other interjurisdictional issues of the last half of the twentieth century, such as water, flood control, roads, and schools, have been more easily resolved. County and regional multijurisdiction entities such as the Clark County School District and the Regional Transportation Commission of Southern Nevada are now the rule.

The need to get tourists to resorts and locals to and from their jobs brought transportation issues to the forefront, especially in the 1960s.

McCarran Airport (later renamed McCarran International Airport) reopened in 1963, upgraded from its initial smaller status to accommodate jet air traffic. At about the same time the federal government lent its support to widen Highway 95 to four lanes along the fifty-eight-mile commute to Mercury, Nevada, and the Test Site. A more ambitious project was Interstate 15, which paralleled Highway 91 (Las Vegas Boulevard) through the city and extended south to Los Angeles. Wrangling over interchange locations, rights-of-way, and project funding delayed its completion until the early 1970s. Particularly cumbersome was the problem of where to place the new road through the already built-up city center. The solution was comparable to those chosen in other cities. Rejecting plans to send the highway through white suburbs on the east or far west edges of town, planners chose instead to pour concrete through the largely black West Las Vegas, the historic location of McWilliams's townsite.

The placement of the freeway through West Las Vegas was one of several sources of racial tension felt in the city during the 1960s. Blatant discrimination was a fairly new local phenomenon at that time. For example, the prejudice against "Okies" for which California is so well known was largely absent in Las Vegas when many of these jobless midwesterners showed up in town hoping for a job at Boulder Dam. African Americans had experienced segregation in their work, living, and leisure space for decades, but black residents (many of whom came from the South) recognized that restrictions here were bearable compared to other parts of the country. As the city grew into a popular tourist destination, however, racially dictated restrictions became more severe. Visitors expected the version of Jim Crow they were used to back home, and resorts began banning black patrons from casinos on Las Vegas Boulevard and Fremont Street starting in 1947. In fact, black performers, such as Lena Horne, Nat King Cole, and Sammy Davis Jr., were not allowed to stay at the Strip properties in which they performed, and were instead forced to West Las Vegas. Racism had become so prevalent that the city acquired the nickname of "the Mississippi of the West."

A threat from the NAACP to march down the Strip in 1960 initiated the first steps toward integration, because the resort owners did not want their "fun and fancy free" image tarnished by television reports of violence and protest. Still, even after initial concessions, minorities were not permitted in dealer, bartender, or manager jobs in casinos until 1971 when a US District Court ruling forced resorts and the unions to make accommodations. That same year residential discrimination practices also were banned. Segregation in the valley's education system brought violence at local high schools in 1969 and 1970. The "solution" was to integrate elementary schools by busing inner-city children to suburban schools for first through fifth grades, and converting inner-city elementary schools to sixth-grade centers where suburban kids would be bused for that single year. By the end of the decade, the situation for African Americans had much improved, but the busing program continued into the 1990s.

Along with the challenges created by growth, the community has reaped many benefits. A dearth in cultural amenities in the valley ended in the 1970s, for example, with the birth of several arts programs sponsored by the University of Nevada, Las Vegas, and Clark County Community College (later Community College of Southern Nevada and then College of Southern Nevada in 2007). Sports lovers have enjoyed the Las Vegas Stars AAA baseball club (now the 51s, after the supersecret military base, Area 51, north of the Las Vegas Valley), which made Cashman Field its home in 1983. Completion of the Thomas & Mack Center on UNLV's campus in 1983 gave a home to the Runnin' Rebels basketball team. Jerry "Tark the Shark" Tarkanian led this squad to an NCAA national championship in 1990, and the team remains a favorite for locals.

The ever-changing Las Vegas Strip provides untold entertainment options for residents willing to brave the tourist crowd. The opening of the Mirage in 1989 paved the way for other megaresorts such as Excalibur (1990), MGM Grand (1993), Luxor (1993), New York–New York (1997), Bellagio (1998), Venetian (1999), Wynn Las Vegas (2006), Palazzo (2008), Encore (2008), City-Center (2009), and Cosmopolitan (2010). More recently, redevelopment in the downtown area, with a fledgling arts district, urban living options, and locally oriented entertainment, aims to bring locals back to the historic core. And, as the suburbs have pushed their sprawling limits to the mountains, local/neighborhood casinos have followed. These Strip-style casino-resort complexes cater to a residential clientele and provide a wide variety of gambling and entertainment choices throughout the valley's periphery.

THE BIG PICTURE

While this overview of the city's history is a mere starting point in a discussion of life in Las Vegas, two major themes from this chapter—opportunity and transition—will surface again and again as local personality traits. For example, the city would not exist today without leaders who seized opportunities provided by external forces. The valley's early residents took advantage of life-giving springs, and this same water gave birth to a small railroad town. Yet, without importing additional water into the valley from the reservoir of Lake Mead, the town would not have survived past the 1940s and certainly could not support today's population.

Owing to its lack of water, Las Vegas also lacks the agricultural or substantial mineral resources that helped support other cities in the American Southwest. Instead, the railroad played this role as a link to outside monetary resources, and was a lifeline that brought goods and services into the isolated desert valley. Geographic luck also played a huge role, especially in the 1930s when the fortuitous proximity to the golden egg of the Boulder Dam project combined with the legalization of gambling in Nevada to allow a tourist industry to begin. Adding to the government influx of money through the dam and other New Deal projects, the war brought another boost to the local economy. Outside investors in hotel-resorts, and eventually mob characters, were the next source of external capital during the postwar era, making Las Vegas a global tourist destination. Most recently, residents and local leaders have grappled with a new and damaging external force in the form of a nationwide home foreclosure crisis and severe recession.

As Las Vegas has experienced extended rapid growth at the hands of external forces, it has had to change to meet new challenges, not unlike other growing urban areas. The Las Vegas Strip is famous for shedding an old image in favor of one newer, bigger, and more elaborate. Likewise, city leaders and residents have modified the trajectory of the local community as opportunities have arisen. Seeing great potential in tourism, the city embraced its frontier image to create an atmosphere appealing to thousands of Americans who traveled through the area to view the dam. Recognizing the unique opportunity in gambling tourism, the region then shifted its focus. At the same time, valley residents, who learned from the mistake of relying too heavily on the single lifeline, worked intensely to maintain ties to industrial and military resources. Today, the story is the same. As the city greets its second century of life and population in Southern Nevada reaches two million, locals continue to adapt. Their continued goal: a culturally mature, ethnically diverse, and economically stable community that will last into the next century and beyond.

A Place in the Desert

Driving south on Interstate 15, five hours from Salt Lake City, a family finds themselves in a barren desert at dusk. Craggy mountains, copper colored in the sunset, surround a valley covered by flesh-colored sand and speckled by sagebrush, Joshua trees, and yucca. They pass signs of civilization—lights from an Indian reservation fireworks stand, a gypsum plant, and a railroad overpass. Having traveled this route only once before in the light of day, their excitement and apprehension grow as they crest a rise near Apex, Nevada. Into view comes a blanket of illumination that seems to move across the adjacent valley's surface as if it were a colony of millions of multicolored fireflies. This vibrant life-form spread across the windshield is Las Vegas, the family's new home.

My favorite part about going home is that moment when I enter the Las Vegas Valley. Every time I do so, the valley lights remind me of the day my family moved to the city more than twenty years ago, an event I reconstruct above. When I moved back to the city in December 2006 to begin fieldwork, my children were not all that different in age from my siblings and me two decades back. It was slightly later in the evening and we entered the valley from the southeast over Railroad Pass. The lights are truly an amazing sight. My then five-year-old son described the scene well, saying, "it looks like a giant golden river." Even in the daytime, as I roll along the city's slightly elevated freeways, the valley's landscape is striking, majestic purple mountains surrounding it on all sides.

At such times, I have wondered: What scene greeted the settlers who entered the valley nearly two hundred years ago after their long journey across the desert? What if we removed all the hotels, highways, and houses that now cover the valley floor? What physical conditions did settlers face as they tried to scrape out a living on the desert soil? In order to answer such questions, we need to discuss the city's geographical site, a term denoting a locale's "specific" position on the earth's surface, its geology, geomorphology, and climate characteristics.

Las Vegas's site does not resemble that of most great cities of the world. No coastline, confluence of rivers, or seaport exists here. In fact, the springs that made the valley barely habitable for early settlers have not been adequate to sustain the population for more than half a century now. The dry, hot summers make living in the area (without air conditioning) nearly unbearable for much of the year. And local mineral and agricultural resources have never yielded enough products to justify a junction or distribution center for such goods, as is the case with the city's midwestern and western counterparts. Furthermore, Las Vegas's relatively short history has precluded it from becoming a financial or industrial center. Indeed, given its geography, it seems that Las Vegas should not even exist.

Somehow, of course, Las Vegas *does* exist as a major American metropolis despite its physical impediments. And that survival is based on the city's "situation," a companion concept to site that refers to a place's relative location: how it fits within a region, relates to other places, and distributes resources, products, and services to other cities and towns in the vicinity.

Presenting this geographical context will help position the events, stories, and descriptions of place that follow in later chapters. Moreover, such a discussion is crucial because of the role that geography plays in the development of culture. Just as a child's character and personality are influenced by the environment in which she is reared, so too does a place derive much of its culture and character from its specific and relative location.

AN UNLIKELY CITY

Greater Las Vegas lies roughly at the center of Clark County, which forms the southern wedge-shaped tip of Nevada. The county, approximately the size of Massachusetts, is bordered by California to the south and Arizona to the east, with the Colorado River and Lake Mead acting as the boundary for much of the Arizona-Nevada line. As mentioned earlier, the metropolitan area is composed of Henderson, North Las Vegas, and Las Vegas cities as well as considerable other unincorporated urban land within the Las Vegas Valley. The city of Las Vegas is the seat of Clark County, and even though the state capital is located more than four hundred miles to the northwest in Carson City, Clark County holds over 70 percent of the state's population. Southern Nevada is the reason Nevada was the fastest-growing state in the nation in both the 2000 and 2010 census counts.

The site in which Las Vegas has evolved is a bowl-shaped basin surrounded on all sides by mountains. The Las Vegas and Sheep Ranges form

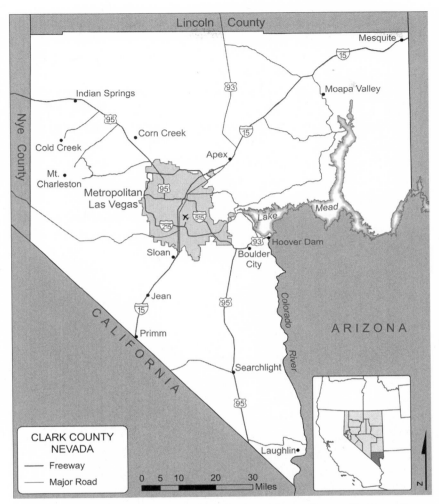

Clark County, Nevada, its major roads, and select cities and towns. Map by author.

part of the barrier to the north, along with the Dry Lake, Desert, and Pintwater Ranges. Frenchman and Sunrise Mountains along with the River Mountains separate Las Vegas from the Colorado River basin to the east, while the towering Spring Mountains mark the western extent of the valley. The southeast extremes of this range, together with the smaller Bird Spring and McCullough Ranges form the valley's southern rim.

Passes between the surrounding mountain walls act as gateways into and out of the valley. Important among these are openings to the northeast at Apex, and to the south at Sloan, which provide passage for travelers along Interstate 15. Northbound drivers on US 95 traverse Railroad Pass in the southeast corner of the valley, and, after passing through the core of the city, exit in the northeast via an elevated, armlike extension of the valley that eventually connects with Indian Springs Valley. Elevations across the valley floor range from two thousand to three thousand feet above sea level and reach nearly twelve thousand feet at Charleston Peak in the Spring Mountains.

Las Vegas Valley is fairly typical of the hundreds of basins that lie between the snaking mountain ranges of the Great Basin. This large physiographic region began a transformation to its present state around seventeen million years ago, when geologic forces stretched and uplifted the earth's crust between the Sierra Nevada and the Wasatch Mountains. As the crust thinned,

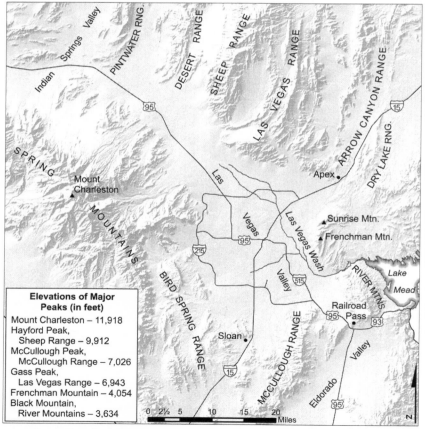

The physical landscape of Southern Nevada. Map by author.

it cracked, and huge chunks of earth shifted, some of them rising and others falling along fault-line fractures. This process, which continues today, created the series of valleys and mountains that characterize the Great Basin landscape. Drivers on north-south Nevada highways have complained about unimaginably long, straight-as-an-arrow stretches within the valleys. East-west travelers, in contrast, remember the ups and downs as they traverse valley after valley, mountain pass after mountain pass. Such is the native landscape on which Las Vegas has been built.

As the mountain ridges and valley troughs formed, other natural processes altered the raw geology. Erosion carved out V-shaped mountain canyons and exposed ancient sandstones, limestones, dolomites, and basalts. Gravity, wind, and intermittent streams carried the sediments downward until they settled on the valley floors, adding to the finer silt and clay deposits from an earlier geologic era when the land was covered by lakes. As the erosion-deposition cycle repeated itself, thousands of feet of sediments covered the original sunken bedrock crust of the valleys. As geologist Bill Fiero interpreted the scene: "Hills of rock sticking above the gravels of valley floors are like icebergs—only the top few hundred feet are exposed. In reality, the small rock outcropping may be the peak of a mountain range rising thousands of feet above its bedrock base."[1] Lone Mountain and Exploration Peak, two isolated summits near the northwest and southwest (respectively) edges of Las Vegas Valley, fit such a description.

Aside from isolated peaks and hills, deposition has given the valley floor a largely flat topography. The Strip and downtown tourist corridor lie roughly at the center of the bowl-shaped depression, and the terrain follows a gentle upward trend from there toward the mountains. Some geologically recent faulting, and subsequent erosion, occasionally exposes bedrock and forms minor hills and escarpments. Drivers along US 95 will notice a prominent rise of Whitney Mesa near the Russell Road exit; Cashman Field downtown sits on a slight rise that provides a beautiful view of both Sunrise and Frenchman Mountains; and Decatur Boulevard follows a north-south linear rise roughly three miles west of the Strip.

On the margins of the valley, at the point where the slope turns from a gentle rise to a steep climb, erosion and deposition have created another characteristic landform of the Great Basin—alluvial fans. These are the deltas of the desert. As an intermittent mountain stream moves through its canyon course, the steep grade and narrow confinement create sufficient velocity to carry sediment, despite a relatively low volume of water. Then at the canyon's mouth the stream is no longer restricted and so loses velocity and drops its bed load. During one spring runoff, that stream may deposit sediments on the left flanks of the canyon mouth, the next year on the right, and a year later somewhere in between. After hundreds of years this process creates a fan-shaped feature made of unconsolidated sands and gravels. In most valleys in the region, such fans have grown large enough to coalesce with adjacent ones to form alluvial aprons. Many of the suburban communities at the valley's edge, such as Summerlin, Rhodes Ranch, and Lone Mountain, are built on these features.

After depositing the sediments that today continue to build the alluvial fans, the water flowing out of the canyons quickly disappears. Much evaporates into the desert air, but some seeps into the unconsolidated sediments of the fans and settles under the valley floor to create underground reservoirs. Historically, this renewable resource made the valleys of the Great Basin important sites for human settlement. Ranchers use this groundwater to sustain their families and herds. Many aquifers hold so much water, in fact, that when tapped, they flow freely. The most famous such artesian flow is Big Springs near the center of the valley, which has served area inhabitants for centuries and, as previously mentioned, gave the city its name.

Residents relied exclusively on the springs and aquifers until around 1970, when usage began to outpace groundwater replenishment rates. Today discharge from wells has diminished to a trickle. Nearly all of the water used by Las Vegans today is imported from Lake Mead through large "straws," a colloquial term for the pipelines that transport water from reservoir to city. Continued meteoric growth in Southern Nevada now is beginning to outstrip even this source (an annual allotment of 300,000 acre-feet of water granted under the Colorado River Compact of 1922). As a result, the Southern Nevada Water Authority (SNWA) is looking for resources elsewhere. This regional government unit charged with ensuring an adequate water supply for Las Vegas has purchased millions of dollars in water rights from Great Basin ranchers to the north and plans to import this groundwater through a series of long-distance pipelines.

Most of the mountains flanking Southern Nevada's valleys follow the characteristic north-south trend of Great Basin ranges. As Bill Fiero has written: "Many years ago a geologist compared [the ranges] to dark fuzzy caterpillars crawling north. Seen from space, these dark tree-clad mountains indeed match the description." But, along the western edges of Nevada extending southward to Clark County, a slight deviation occurs, yielding "discontinuous and arcuate ranges. . . . The confused ranges look as if a fifty-mile-wide eggbeater had moved through plastic rock."[2] Geologists call this eggbeater the Walker Belt, or Walker Lane, a series of lateral faults that they believe have caused earthen displacements of up to 120 miles. The Spring Mountains in their southeast-northwest trend are part of this deviation.

The mountain ranges surrounding the Las Vegas Valley also differ from their counterparts in the Great Basin by having less vegetation, particularly those ranges where elevations fail to reach five thousand feet. Rather, Southern Nevada mountains generally—a major exception being the taller Spring Mountains—resemble their bare, copper-colored regional counterparts in the desert Southwest.

The Las Vegas Valley's most striking deviation from the Great Basin pattern, however, is its hydrology. Bounded on the west by the Sierra Nevada, on the east and south by the Wasatch Mountains, and on the north by the ridges of the Snake River Plain of southern Oregon and Idaho, the Great Basin is defined as a place from which water never escapes the valley, except into the air through evaporation. Stated differently, any water that accumulates within a true basin will not flow to an ocean outlet. Nevada's Humboldt River is a good example of such a watercourse. The Great Salt Lake is another, a remnant of the Pleistocene Lake Bonneville. Lake Mead, on the other hand, does have exterior drainage. Filled by the Colorado, Muddy, and Virgin Rivers, water held by the lake flows out of Hoover Dam and continues down the Colorado toward the Gulf of California.

So, if one adheres to strict hydrologic definitions, neither Las Vegas nor the majority of Clark County lies within the Great Basin. Most scholars choose instead to include Southern Nevada within a larger Basin and Range physiographic province, taking in not only the Great Basin proper, but also geologically similar portions of Nevada, California, Arizona, New Mexico, Texas, and Mexico. The Colorado River cuts a deep incision in the belly of the Basin and Range province, gathering volume in Clark County from major tributaries such as the Muddy and the Virgin. The Muddy's origins are north of Las Vegas at the convergence of several spring-fed streams. The Virgin begins in southern Utah, entering the Silver State via Arizona, near Mesquite. As these tributaries continue the process of headward erosion today, they move the borders of the Great Basin back farther and farther.

One of the minor Colorado tributaries, the Las Vegas Wash, is of particular local importance. Fed historically by free-flowing springs in the valley, this wash makes its way to the Colorado River by way of a gap between

These concrete arroyos, like their natural counterparts, are wet only seasonally. But when the periodically torrential downpours arrive, waterways like this one near Fort Apache and Sunset Roads help to keep the flash floods off the roads and out of the neighborhoods. They channel runoff toward detention basins and eventually to the Las Vegas Wash and Lake Mead. Photo by author, March 2007.

Frenchman Mountain and the appropriately named River Mountains.[3] Those springs no longer flow, but today the Las Vegas Wash transports runoff from cloudbursts and excess lawn watering. It also takes on effluent treated and returned to the environment under the direction of the Clark County Reclamation District. The SNWA has arranged that all such water returned to Lake Mead can be exchanged for additional water allocation credits above the 300,000 acre-feet baseline. With this drainage feature, I prefer to compare the Las Vegas Valley to a bathtub. Its elliptical shape is oriented northwest to southeast, tilted slightly, with the Las Vegas Wash acting as a drain at its southeastern rim.

A DESERT CITY

Las Vegas has a midlatitude desert climate and is classified vegetationally as part of the Mojave Desert. New residents of Las Vegas often complain about the lack of four distinct seasons. This perception has some truth to it, as evidenced by temperature patterns. Only short intervals in April/May and again in October/November show moderate spring and fall conditions between more extreme and longer-lasting periods of winter and summer when artificial heating or cooling is needed for local homes and businesses.

The towering rain shadow of the Sierra Nevada spreads across southern Nevada. Within the valley, average annual rainfall is less than 4.5 inches, and gauges in 2007, a particularly dry year, recorded only 2.7 inches. Most precipitation falls in the winter between January and March, a result of occasional storms from the Pacific. Moisture-rich monsoonal air from the south sometimes pushes its way as far north as Las Vegas in the summer months, providing another period of heavier-than-normal rainfall. Many locals enjoy the occasional venture out to the desert, away from buildings and lights, to view the brilliant lightning that accompanies these summer storms.

Rainstorms can be a hazard in a desert city. Every year, despite warnings plastered across billboards, flash floods inevitably catch drivers off guard. Footage of cars floating abandoned along one of the city's flooded thoroughfares regularly makes the local evening news. The Clark County Regional Flood Control District has worked to relieve some of the pressure from these summer storms. For example, the notorious inundation of Charleston Boulevard's railroad underpass, remembered by longtime locals, no longer occurs. Still, floods continue to damage homes, take lives, and deposit huge amounts of sediment along valley roads. On a morning following one such storm, I drove over several inches of debris on West Flamingo Road as I left my neighborhood. The bed of sediment included rocks four or five inches in diameter.

Snow, on the other hand, is rare on the valley floor. Possibly once a year Las Vegans will see flurries, but accumulations typically occur only twice a decade, and melt away fast. Greater accumulations, of course, are found in the mountain areas; nearby Mount Charleston even has a modest ski and snowboard park. Other potential threats to health and property include earthquakes and occasional wind and dust storms.

Amid such hazards, sun and warmth are the greatest advantages many Las Vegans see to living in their Sunbelt city. Winter temperatures are mild; the official thermometer at McCarran International Airport rarely dips below the freezing point. Park playgrounds that would sit dormant in most midwestern towns between December and March are alive with activity in the valley. An observer would find less activity in those parks as temperatures reach their summer extremes. The official all-time local high was 117 degrees Fahrenheit, recorded in July 2005, but during the summer of 2007, local news outlets regularly reported readings over 120 degrees at stations in the lower elevations of the valley. Even if the official daily temperature didn't

TOP: Consequences of a summertime flash flood. Transported rocks and debris (including an abandoned shopping cart) from an undeveloped lot hinder access to a gas station at Tropicana Avenue and Fort Apache Road. Vehicle accidents, such as the one visible in the background, also are common occurrences when such storms occur. Photo by author, August 2007.

BOTTOM: Runoff from a summer storm. Water flows through a pedestrian walking path near Patrick Lane and Fort Apache Road, which doubles as a drainage channel. This type of flash flooding can be hazardous to life and property. In the lower right of the image is a barrier intended to divert the flow from construction on Fort Apache. Photo by author, August 2007.

break the record, the mean annual temperature for 2007 was 71.1 degrees, one full degree above the previous record set in 2003, and on eighty-eight different days the temperature soared above 100 degrees.

Still, praise outweighs the complaints for Las Vegas's climate. After all, locals enjoy, on average, more than 85 percent of their days in the sunshine. In addition, low humidity levels (around 20 percent at midday in July) moderate the effect of summertime heat. This climatological fact has actually spawned a local joke of sorts. To get through the summer, many locals recite the well-worn "It's a dry heat," as they rush from air-conditioned cars to air-conditioned offices and homes. But, when it gets really hot, it's hot. I tend to agree with *Las Vegas Review-Journal* publisher Sherman Frederick as he ranted: "From now on, when the temperature climbs to 103 degrees or higher, I'm not telling the 'dry heat' lie anymore.... Instead, if a guy from Iowa says, 'Jeez, it's hot in Vegas,' I'm going to reply: 'No kidding! You must be the president of Dubuque's Mensa chapter. Now get back inside the casino before you burn to a crisp.'"[4]

Given the climate and environment, the cards seem stacked against a city of two million within the southern Nevada desert. Yet, local people have made living in the desert possible, even enjoyable. Air-conditioning is a must, of course, and the school district also places awnings over playgrounds so the students can safely have their outdoor recess when the temperatures climb. Access to water in nearby Lake Mead helps too. But, as the city grows, many people have asked, "What happens when Las Vegas runs out of water?" I am certain that, given the city's ability to adapt in the past, Vegas will find a way.

A POSTINDUSTRIAL METROPOLIS

Just as Las Vegas people have adapted to harsh environmental circumstances, they have done the same in response to an economically unfavorable geographic situation. The city lacks many of the ingredients that promoted economic growth and regional importance in other major US cities. Compare, for example, the role within region and country of three such communities. New York was (and is) the hub of immigration to the United States, its protected harbor the gateway to freedom and liberty for many new Americans. Chicago, also a gateway city, pioneered railroad construction to the east and west, thus becoming the market center for the agricultural wealth of the Midwest. As the country's western doorstep, San Francisco built its foundations on mining wealth and the first transcontinental railroad. As well-connected transportation centers, each of these cities became major entrepôts to a vast hinterland.

Las Vegas has never been an immigrant capital, financial hub, or market center for the nation or even the western United States. Nor did the city have the agricultural, mining, manufacturing, industrial, or military significance of its western competitors in Albuquerque, Phoenix, Tucson, and Los Angeles. Las Vegas, for most of its history, has been nothing more than a way station and watering hole for nomadic hunter-gatherer groups, weary wagon travelers, and, later, thirsty railroad passengers.

The tiny railroad town remained of little consequence, a speck on the map, until an influx of government-funded projects provided a boost in the 1930s and 1940s. Construction on Boulder Dam in the early 1930s brought jobs and tourists during a time of depression, and thereby saved the whistle-stop from a ghost-town fate. Power from the dam and water from the lake it held back made it possible to create an industrial center southeast of Las Vegas in what became Henderson, and plants there provided needed goods for Allied involvement in World War II. Also born during the war was Nellis Air Force Base, which turned out thousands of pilots for the European and Pacific fronts. But, the economic impact of such enterprises did not compare with what other southwestern cities experienced. Phoenix, for example, hosted four military bases during the war along with several civilian-operated aerospace companies, and this continued for many decades.

Anyway a person looks at it, Las Vegas's raison d'être, its lifeblood, its major export to the world, is casino gambling and resort entertainment. In fact, the rise of the resort business, casinos, and associated vice in the 1940s and 1950s was partly responsible for the city's failure to become an industrial center.[5] Yet, Sin City sold a product that was appealing to many people in postwar America, and the fact that it was away in the desert made it that much more attractive. The Las Vegas of this era was the only American city that embraced and legally marketed to tourists who desired to act out fantasies that were largely socially unacceptable back home. As American ideals shifted gears starting in the 1960s—when the vice sold in Las Vegas was seen as a valid form of entertainment, and a Vegas weekend no longer elicited a wink or a leer—Las Vegas became more culturally acceptable.

By the 1990s, Strip resorts began marketing to families. A teenager at the time, I felt these winds of change firsthand. My first high school job, in 1993, was as a go-kart track attendant at Funtazmic, a family recreation center just a mile or so west of the Strip. As we prepared for the grand opening, the owner gave his new teenaged employees a pep talk. He said, in effect, "This is the perfect time for something like Funtazmic. Las Vegas is becoming a place where parents are bringing their children and desire family-oriented entertainment options. MGM Grand has their theme park and other resorts have roller coasters. Funtazmic will be part of this *new* Vegas market."

My boss was right, at least partially. Families came. The Mirage had a dolphin exhibit, New York–New York had a Coney Island-themed roller-coaster, the Bellagio had a $350 million gallery of fine art, Circus Circus opened an all-indoor Grand Slam Canyon Adventuredome complete with double-loop roller-coaster, and the list goes on and on. But the family-oriented Las Vegas did not last for long. Again adjusting to fit market needs, resort leaders (as they always will) have now shifted focus away from the family back to adults and adult-centered entertainment. Ironically, Funtazmic closed within three years of its grand opening, presumably at the hands of this same Vegas culture change.

As a cultural and moral shift took place in the United States in the 1990s, Las Vegas became openly acceptable. Shady gambling became corporate gaming, and Southern Nevada became an entertainment destination, offering much more than smoke-filled casinos. Today, Las Vegas's export is a postindustrial one: entertainment for the masses. If the role of Las Vegas is that of an entertainment consumerist religious hearth, then perhaps we should compare Sin City's geographical situation not to that of Phoenix or Los Angeles, but to that of Salt Lake City, the other religious hearth in the West. The numbers show that Las Vegas has indeed become a tourist mecca. In 1970, just under seven million people visited Southern Nevada, according to the Las Vegas Convention and Visitors Authority (LVCVA). That number more than doubled fifteen years later. Then, with the opening of the Mirage in 1989 and the megaresort explosion that followed, visitor volume topped 29 million by 1995. After reaching a high point in 2007 of more than 39 million tourists, numbers declined with the onset of global recession to 36.3 million in 2009, but then rebounded to 38.9 million in 2011. These totals put the city far ahead of the real Mecca, which hosted around 2.9 million religious pilgrims in 2011, according to the Saudi Arabian Central Department of Statistics and Information.[6] Even during the lull of 2009, Las Vegas was the sixth-most-visited American city, according to Forbes.com, after Orlando, New York City, Anaheim, Chicago, and Miami. Furthermore, the Strip maintained the number two spot among US tourist attractions (behind Times Square), and hotel occupancy rates in Las Vegas remained higher than in any other tourist market in the country.[7]

Room occupancy rates are a statistic often touted by city boosters. Historically, LVCVA has reported, Vegas hotels maintain annual occupancy rates of 90 percent or higher, with dips below that threshold during temporary slumps. A key to this success is conventions, which brought in more than 6.3 million delegates in 2006 alone, resulting in a nongaming economic impact of more than $8 billion. Add that to nearly $10 billion in gaming revenue for the same year and you get a good picture of the impact that casino-resorts have on the Southern Nevada economy.

Tourism also fuels growth off the Las Vegas Strip. It takes a huge workforce to maintain slot machines, valet cars, clean hotel rooms, set up and take down conventions, and shuttle hordes of tourists. The *2007 Las Vegas Perspective* reported that nearly 30 percent of the valley's nonagricultural workforce was employed in leisure and hospitality services. That is more than double the combined number of jobs in mining, construction, and manufacturing industries. In addition, 19 percent of the city's gross domestic product derives from such employment, while visitor spending accounts for an additional 29 percent, making Las Vegas more reliant on leisure and hospitality revenue than other US cities and the nation as a whole. Furthermore, fully 85.5 percent of the valley's working population is employed in the broader service economy.[8] Indeed, Las Vegas is a shining example of a postmodern, postindustrial metropolis.

Despite the ever-powerful influence of tourism, labeling Las Vegas a one-industry town is not totally accurate. After all, that 30 percent of the working population needs doctors, grocery stores, malls, and auto shops. New buildings on the Strip and in the ever-expanding suburban neighborhoods have nurtured a strong construction industry in the valley. In addition, the cultural sanctioning of the Sin City image plus the city's burgeoning population, Nevada's friendly tax structure—it has no personal or corporate income tax, no gift or inheritance tax, no inventory tax, relatively low estate tax, and low employer payroll tax—and a low cost of living relative to other attractive climes of the Sunbelt all have combined to make the city an attractive destination for major corporations. One of the earliest such companies to open an office in the valley was Citigroup, which maintains a large call center that employs more than two thousand workers. In more recent years Zappos has become a symbol of corporate success in the city. The online shoe and clothing retailer has affixed its footprint in downtown Las Vegas and seeks to transform the city's core from a largely casino and city government hub into a center of business, retail, urban living, and local entertainment.

LINKS ACROSS THE DESERT

Transportation linkages are important in the geographic situation of a place. In Las Vegas, such linkages are crucial, given its relatively isolated location in the Mojave Desert. Interstate 15 brings weekend revelers from Utah and Southern California. The traffic is such a problem, in fact, that Californians who do not wish to spend eight or ten hours on what should be a five-hour trip avoid travel north to Las Vegas on I-15 on Friday and south toward home on Sunday. A former resident of Orange, California, explained how he frequently passed through Las Vegas on his way to southern Utah and Idaho. Several years ago he was stuck in that I-15 gridlock so long that he was able to get out of his vehicle and carry on a lengthy conversation with his highway "neighbors." Las Vegans who take weekend vacations to Southern California often grin at the seemingly endless line of headlights across the median, glad to be going the opposite direction.

Cutting through the valley from northwest to southeast is US 95, a highway locals refer to as the "Expressway." This artery lacks tourist traffic from the north (445 miles of barren highway separate Las Vegas and Reno, the next city of any size in the state), but it brings many travelers into the city from Arizona. Some northbound drivers choose Highway 93, which connects with 95 at Boulder City after crossing Hoover Dam. The latter path became the quicker alternative in 2010, when drivers gained the option of using the newly completed Hoover Dam Bypass to avoid what used to be an hours-long bottleneck at the dam. This bridge, formally known as the Mike O'Callaghan–Pat Tillman Memorial Bridge, spans Black Canyon sixteen hundred feet downriver from the dam and nine hundred feet above the river's flow. It significantly cuts travel time for Phoenicians (Phoenix residents) headed north with a four-lane freeway for most of the stretch between Sun City and Sin City. In recent years, many local boosters, planners, and politicians have pushed for the incorporation of this corridor into the interstate system, designated as Interstate 11. As a hub for this improved roadway, which is expected to extend from Phoenix through Nevada all the way to Oregon, Las Vegas would likely reap great economic benefit.

Notwithstanding its convenient highway linkages, Las Vegas could not maintain its current existence without McCarran International Airport. In

2004, passenger traffic passed the forty million mark, most of them tourists. Passenger volume reached more than forty-seven million in 2007 before dipping again to around forty million with the onset of the nationwide recession.[9] Las Vegas locals, of course, find great advantage in having so many flights in and out of town. I myself have enjoyed this convenience, and many of my interviewees cited it as one of their favorite parts of living in the city. Historically, city leaders have marketed this same trait, adding to it the city's proximity to the natural and cultural beauty of the West.

The fifth busiest airport in the country and tenth in the world in 2006, McCarran is beginning to outstretch its capacity. The Clark County Department of Aviation opened a third terminal in 2012, and some smaller plane traffic is often diverted to municipal airports in North Las Vegas and Henderson. The county also is making plans for another potential airport in the Ivanpah Valley, thirty miles south of McCarran near Primm, Nevada, and the California state line. Ivanpah Airport would be expected to serve another sixteen million passengers.

In addition to transporting locals and tourists, road and air linkages provide important corridors for the movement of goods throughout the region. McCarran officials, in fact, market their facilities for freight transport. An absence of an inventory tax and Las Vegas's central location in the West also have helped to make the city a cargo distribution and warehousing center within the region. In 2006, 201.7 million pounds of cargo passed through the local air cargo center, located on an eighty-acre Foreign Trade Zone at the airport. Many of the more than twenty-five transportation companies that use the cargo center are passenger airlines, but FedEx and UPS carry the bulk of the load.

Interstate 15 and Highways 93 and 95 also play important roles in cargo transportation. One of the forces behind the $234 million Hoover Dam Bypass is Highway 93's role as one of the North American Free Trade Agreement (NAFTA) routes. Tractor-trailers are not allowed over Hoover Dam and were forced to travel to Las Vegas by longer alternate routes until the bridge was completed. Now that it's open, thousands of trucks use the quicker route over the bypass everyday.

ONE CITY, MANY REGIONS

A discussion of location also should address how the place fits into regional culture. Such divisions, although somewhat arbitrary, are essential for both self-identity and how outsiders see and evaluate the city. Four such regional labels are important for Las Vegas.

The city, first, is firmly a part of the West. Even laying aside its latitude and longitude coordinates, the personality of the city conforms to either mythical or factual descriptions of that region. The city's birth, for example, came at a relatively recent time in the country's history. In fact, its 1905 date is considered late even for the West. Partly because of its belated development and partly because of its isolated location, Las Vegas until the 1930s maintained a classic western gun-slinging, cowboy, libertarian image. Some of the characteristics hold true today, especially the way the city continues to hold out the possibility of "treasure." Visitors and transients no longer seek the silver and gold of the Comstock Lode or Sutter's Mill, but rather a slot jackpot, a winning hand in a Texas Hold'em tournament, or an opportunity for a new life and a new beginning.

Modern-day Las Vegas is just as easily placed within the American Southwest. The physical landscape surrounding the city is one indication, with its copper and purple hues, its unvegetated mountains, plateaus, and buttes, and its Joshua trees and yucca. A dry climate is further reason for such placement and demographics more still.

In the last twenty years, Las Vegas has experienced a huge influx of Latino/Latina immigrants, adding a cultural flavor that has long been part of life in Phoenix, Albuquerque, and San Diego. Between the 1990 and 2000 censuses, the Hispanic population in Clark County exploded from 85,000 to 300,000. In 2010, the Census Bureau reported that more than 568,000 Hispanics lived in the city, which was 27 percent of the total population. By November 2006, the number of Hispanic students enrolled in the Clark County School District outnumbered white students for the first time.[10] The northeastern quadrant of the city (directly east of downtown) now resembles South Phoenix and East Los Angeles in many ways. But this experience of Las Vegas Hispanics is atypical of other cities in the region. The city gives immigrants more of an opportunity to "make it," earning wages and benefits comparable to many typical middle-class Americans.

Another migration trail places Las Vegas within the Sunbelt region of the United States. Stretching across the southern third of the country, this region is defined by its enormous numbers of new residents from colder climes. Clark County, of course, epitomizes this movement with its attractive climate, recreational and entertainment opportunities, affordable cost of living, and fast-growing population. Retirees are the most visible component of Sunbelt culture, coming to Vegas in search of golf courses, warmer temperatures, fewer taxes on their fixed incomes, and a cheaper cost of living than Florida or California.

Las Vegas also falls within the borders of a fourth region, one less conspicuous and less obvious than the previous three. This one roughly follows the borders of Nevada, but calling the *region* "Nevada" does not capture the essence of the geographical character that is felt by its residents. Instead, I prefer the moniker provided by a book of essays about the personality, unique spirit, and mystique of the state: *East of Eden, West of Zion*. The name comes from an analysis by William D. Rowley,[11] a historian at the University of Nevada, Reno, who wrote, "The land between California and Salt Lake City was a virtual Land of Nod—east of Eden and west of Zion's religious settlements. It was held in low regard as a place for serious settlement, even unfit for habitation, possibly a place of exile."[12]

Many times it is easier to describe a place by explaining what it is not, and referring to Nevada as east of Eden and west of Zion does just that. At the same time, it also suggests (as does the book with that title) the vast differences between subregions of the West. Nevada has stood on its own, more than most states, making its own character, and providing a place of exile and opportunity for its people.

Las Vegas carries with it much of the Nevada personality locked up in the phrases describing this place in between. Like a confluence of two powerful cultural rivers, one with headwaters in Salt Lake City and the other originating in Los Angeles, the Las Vegas landscape and character is a product of both. Southern California and Utah transplants to the southern Nevada desert have always been numerous and bring with them architectural tastes and traffic congestion from the former locale and religious and moral conservatism from the latter.

A refuge for some and a trail to success for others, Las Vegas has simultaneously provided a place of exile and opportunity for those seeking a new life. As I will demonstrate in the chapters to follow, Las Vegas's ability to be a place both of new beginning and last resort is one of its most defining characteristics. The city is often referred to as a land of opportunity, a "can-do" town. I believe its "East of Eden, West of Zion" origins explain much of that character. It is the in-between place that gleans what it can from those places that have a well-defined identity and provides something other places cannot (or will not) in order to forge its own personality.

THREE

Watch 'Em Come, See 'Em Go

In 1872 a huge gold vein was discovered in the Snake Range of central Nevada near the Utah border. The resulting mining town of Osceola reflected the experience of many of its Nevada siblings: it boomed, attracted thousands of treasure seekers, produced millions in wealth, and then busted. All that remains in Osceola is a memory preserved in ruins, the occasional rancher keeping watch on sheep or cattle that roam the surrounding slopes, and a state-sponsored historical sign along lonely Highway 50. The sign reminds the curious passerby that Osceola, like many other Nevada places, faded out of existence once the gold did. Major mining operations ceased there around 1940.

A four-and-a-half-hour drive south of Osceola, another sign tells a story of a different kind of boomtown. Designed by Betty Willis and installed in 1959 by Western Neon, the WELCOME TO FABULOUS LAS VEGAS sign marks the former entry point to the city for visitors traveling north from California along Highway 91, then known as the Los Angeles Highway. Today, it is a photo opportunity for hordes of tourists, who now come by way of Interstate 15 or McCarran Airport and want to commemorate their trip to the city. The sign symbolizes a "fabulous" weekend for such visitors, a short escape from the monotony of a commute, office work, and community rules that are (legally) breakable in Las Vegas. Most of all, the images of silver dollars beneath the letters W-E-L-C-O-M-E sing a new version of the old siren song of Nevada's mines: a chance to hit it big. Willis intentionally never copyrighted her design, later explaining "It's my gift to the city," so it has been copied in pins, paperweights, T-shirts, and a thousand other items of Vegas kitsch. Indeed, this sign has become the epitome of the Vegas image perceived by the world.[1]

The sign also speaks volumes about the local's experience in the city. To be certain, residents typically do not park in front of the now-closed Klondike Casino, daringly jaywalk across multiple lanes of traffic to the median of Las Vegas Boulevard, and snap pictures of the famed neon sculpture. Instead, the

The world-famous WELCOME sign. This oft-photographed feature symbolizes the open arms of Las Vegas to both tourists and locals. Photo by author, July 2005.

sign represents a welcoming opportunity for success, a new beginning, and a chance to live the American dream. Some will strike it rich, become successful, and make a life for themselves in the city. Others will come up empty-handed and leave town for the next opportunity.

Las Vegas actually has a lot in common with mining towns like Osceola, even though this city was not founded on the discovery of some mineral or metal. For many decades following Nevada's mining boom, Las Vegas has remained one of America's most attractive boomtowns. This trait is recognizable not only in the modern-day gold fields of the casinos, but also in the city's seemingly limitless opportunities, welcoming attitude, and pervasive entrepreneurial spirit.

Yet, Las Vegas warps the boomtown model in that its "mines" show no signs of depletion. Even with a fracturing national economy that reduced visitation and spending in the city and burst the housing bubble, stunning the local real estate market, new megahotel openings during this time, such as MGM Resorts' CityCenter and its neighbor the Cosmopolitan, provide hope that the market will spring back and that the city's suburbs will continue to creep toward the mountains' edge. And, unlike the sign at Osceola bolted to a lifeless metal grate in the shape of Nevada, Las Vegas's sign still glitters every evening at dusk.

Before turning to several aspects characterizing the city's growth, I want to profile Las Vegas transience. Many people come to the city seeking some version of a rich vein of gold and find it. Others have high hopes but end up without the success they expected. In many ways, the stories of locals coming and going are similar to those of other migrants who choose to move to Anywhere, America: a new job or relocation, a chance to be closer to family, or just a new start. But in other ways, these stories are unique, contributing to a Vegas lore all its own.

THE START-OVER TRANSPLANT

Las Vegas is commonly perceived as a place where one can make a new start on life. For reasons I discussed in the previous chapter (climate, plentiful jobs, relatively low cost of living, friendly tax structure), the city is an attractive destination for such a person. That draw has existed for decades. John Beville wrote of early Las Vegans who came to the desert town to start anew: "The impulse which prompted their coming . . . and staying . . . was simply the inexplicable stubborn cussedness of a pioneer—the urge to create something from nothing."[2]

I interviewed a descendant of one such pioneer, a rare fourth-generation Las Vegan, who told me the story of his great-grandfather, Ed Von Tobel, who was working at a lumberyard in Los Angeles in 1905, the year of the Las Vegas land auction. He got wind through a newspaper ad that a railroad company was giving free one-way train tickets to people who wanted to attend the event, plus an offer for a free return ticket if they completed a transaction. So he and a business partner, Jake Beckley, took the offer and bought adjacent parcels for $100 each. When Ed returned to LA he was promptly fired; his boss had seen him at the auction. A. D. Hopkins wrote of the event: "Until then, Von Tobel hadn't made up his mind to move to Las Vegas. But with no other immediate prospects, he borrowed money from his father to buy lumber, nails, hardware, a delivery wagon, and a team to pull it. . . . Surely a lumberyard would make money fast in a town being built from the ground up."[3] It was not so simple, but, as Von Tobel's great-grandson told me, his was the only one of many early lumberyards (including one owned by Ed's old boss) to endure. In fact, the Von Tobel lumberyard lasted until the 1970s, when it was purchased by a large corporation.

Other interviewees exemplify the Las Vegas character as a place of second chances. Al Zanelli is one such individual. When I asked what brought him to Las Vegas, he said it was to escape a stalking girlfriend. After arriving, he enrolled in a dealers' school and worked at the casino tables for five years before transitioning to air-conditioner repair. He was working in an auto parts store when I met him, having left the A/C business because of fears of falling and a dislike of the summer heat brutally intensified on his clients' rooftops.

Charley Sparks longed to get out of Southern California for some time before his mother's terminal illness provided the impetus. He needed a change, and so in 1992 he closed his barbershop in LA and reopened it in Las Vegas. The operation is still successful today.

Divorce was a push factor mentioned by two interviewees. Jane Willems, having separated from her husband in 2005, sold her home in San Diego and made the move to Las Vegas to be close to her sister. Tina Lewis also came from San Diego with her children in hope of making a new start following her divorce. As was the case in many other transplant experiences, her choice was driven by a lower cost of living relative to Southern California. She said they dug in, got an apartment, eventually moved into a house, joined a church, and became part of a community. Relating her experience to early Las Vegas transplants like Von Tobel, she commented: "People come here from all over the country and the world. There is a spirit of renewal here. You can do whatever you want. People come here who want a fresh start and a different perspective on life. I was in San Diego for thirty-two years and I got here and said, 'Wow! This is like being a pioneer.' You can come here and reinvent yourself. It's affordable and you can renew who you are as a person."

THE PRAGMATIC TRANSPLANT

Another set of locals made Las Vegas home in order to improve their quality of life. The underlying reasons behind their moves are broadly similar to those of the Las Vegas "start-overs," but their driving factor is more simply practicality. In this group we find Professor Rudolf Lusetto, who teaches at the College of Southern Nevada. From Colombia originally, Professor Lusetto came to Las Vegas by way of Chicago, where he and his wife assisted in the care of the wife's mother. The mother-in-law wanted to be closer to her grandchildren, a handful living in Vegas and others in New York. In the end, weather was the deciding factor. Professor Lusetto said they didn't want to be anywhere near snow after enduring it so long in Chicago, especially considering his equatorial roots.

Belinda Kaneshiro also cited family reasons. After the events of September 11, 2001, she realized that family was more important than money and so left "corporate America" and her high-paid job in Washington DC in 2006. She said that Las Vegas is not as expensive as California, allowing her to live in the city on a medical secretary's wages. She also mentioned that the many flights in and out of Las Vegas allow easy visits to her home state of Hawaii. Another Hawaiian interviewee told me that Las Vegas is home to an unusually high number of residents from the Aloha State seeking similar opportunities and enjoys a high volume of Hawaiian tourists. In fact, a downtown hotel-casino called the California caters specifically to such visitors. She said, "Hawaiian locals [in Las Vegas] call this the 9th island."

Retirees often cited a variety of practical reasons for coming to the city. At the February 2007 First Friday event, a monthly art walk and street festival in the downtown Arts District, I met a retired couple from Boston who told me they came to the city months earlier. Asked why they came, the couple said, "Better entertainment, better food, and better weather. One hundred ten degrees is better than negative ten degrees!" The couple added, with a grin, that Boston was getting snow that day and they were not at all disappointed with the slight winter chill of that evening. Another woman I spoke with came to Las Vegas in 2004 from St. Louis, retiring early to be closer to her parents who were living in California. She also commented on the weather, glad to be out of the ice and cold and preferring the heat to scraping snow and ice from her car.

Climate, convenience, and quality of life also figured into Jeff Simmons's decision to move to Las Vegas as a young man in 1979. After completing military school in Florida and a short stint in Palm Springs, California, Jeff was looking for a place that fit his personality. He said, "In Palm Springs, you're either old and retired or poor and on welfare. There's no middle class. That is why I came to Las Vegas." (Then he added, jokingly, "I came here for the cooler climate.") He found what he was looking for, relishing the entertainment scene in the city. He had attended 699 concerts when I met him, and his studio apartment walls were plastered with rock and roll memorabilia collected over the years. He has found steady work as a photographer (his specialties are concerts and Vegas hotel implosions) and has been a dealer in several area casinos.

Since the boom days that accompanied Hoover Dam's construction, Las Vegas has been a place where most anyone can find a job. Referring to a crane business he successfully operated in the city for decades, one longtime local explained: "You do a good job and be on time and you'll get all the work you want in this town."

Such a can-do spirit and entrepreneurial attitude is also reflected in Darren Sedillo's story. When I asked Sedillo what brought him to Las Vegas, he cited the high cost of living in his native Southern California. After Sedillo's sister-in-law moved to Las Vegas's Summerlin suburb in the late 1990s, he and his wife made a habit of visiting once or twice a year. "You know, when you visit Vegas, you want to go to the Strip," Sedillo told me, "but since we were staying with them, it was a little different." He continued, "Pretty soon, as we wandered around, we'd see grocery stores," revealing his surprise at this. "We did some more wandering. One time, we went to a home show, you know where they show new home models. We looked at a three-thousand-square-foot [new] home in [the] Seven Hills [neighborhood] that was $175,000." That was appealing to the couple, especially considering that even a fixer-upper in the California market would cost something like $400,000. "So, we made a five-year plan to save and move out here. We put our deposit on our house in September 2002, when it was just a slab of dirt, and moved [to the Silverado Ranch neighborhood] in 2003." He added another appeal of Las Vegas, this one related to his career, "I knew I wanted to start my own business and Vegas would be a great place" for this. When we met, his pet services business had grown to include several employees and a fleet of trucks and trailers. Sedillo, like so many others, saw opportunities to do more with his life and career in Las Vegas. He confirmed this belief in pointed terms, using a refrain I heard often: "If you can't find a job here, you're either lazy or addicted to something."

THE TOURIST TURNED LOCAL

Sedillo's experience visiting Las Vegas as a tourist before deciding to move hints at the pull factors of the next group of transplants I encountered: tourists turned local. Given the city's insider/outsider character, this group of transplants is particularly important. In fact, I could easily reclassify many of the transplants from other groups into this category because so many newcomers are drawn (directly or indirectly) to Las Vegas because of the Strip. But, even those people who are initially enamored with such amenities eventually ease away from that lifestyle, like Tracy Snow, whom I quoted in the introduction. The tourist-turned-local phenomenon is one that many Las Vegans can relate to, ubiquitous enough that Boyd Gaming, owner of several of the locals-oriented casinos in the valley, recently displayed an advertisement on a Highway 95 billboard near Eastern Avenue that proclaims, "Proudly Serving Locals Since They Were Tourists."

Several stories illustrate this trend. The Donalds, a couple from New York City, had regularly visited Las Vegas since the 1980s. They purchased a time-share condominium in Jockey Towers on Las Vegas Boulevard and began spending two-week vacations there, using the spot as a springboard for travels to other parts of the West. As Mr. Donald neared retirement they began to think about a permanent move, a possibility enhanced by the realization that no family remained in New York. So, on one of their visits, they began to look at homes in suburban retirement communities. They first looked in Summerlin, which Mr. Donald told me was too far away from UNLV's cultural offerings. They then found a place they liked at Sun City McDonald Ranch in the south of the valley. There they enjoy the local arts and culture, including trips to the Arts District and showings of the Metropolitan Opera broadcast to local movie theaters.

I met Matt one afternoon as he walked his dog Ronny in my neighborhood. As was common in many by-chance encounters I had, our conversation turned to where each of us was from and how we came to live in Las Vegas (in itself a telling indicator of the city's transient character). From Pittsburgh originally, Matt came to town by way of Florida. He plays poker and, like Jeff Simmons, enjoys rock concerts. As a result, he visited Las Vegas four or five times a year. Matt's change in perception of the city resonates with Darren Sedillo's experience with grocery stores: "The first time I came here, I said, 'I could never live here. Sure it's a nice place to visit.' . . . It's fine to come to Vegas and have a good time, but . . ." After a few more trips to town he realized, "there's more to Las Vegas than that," and reckoned that his frequent visits justified a permanent move. Matt arrived in town in December 2006. His work in architecture and design has allowed him to continue client service in Florida while transitioning to new ones in Nevada.

Carlin, a Minnesota native and diehard Vikings fan I met in the park as our children played, had visited Vegas each March to party with friends from college, a tradition that began their senior year. As we sat at his kitchen table the day after our original encounter, he told me about his first visit. "It was seventy degrees and beautiful that afternoon," he asserted, "and I knew I would like it here." But, he still had family in Minneapolis/St. Paul, and it took five years to finally make the move in 1998. His first job was at a casino sportsbook, which, he said, was a good fit given his love of sports. A higher paycheck motivated him to become a card dealer, a career track common to many Vegas transplants. He started out at one of the downtown casinos and then moved with his floor manager to a larger, luxury property on the Strip where he can make a nearly six-figure income.

THE VEGAS DREAMER

For the tourist turned local, the draw to Las Vegas is initially gambling and the entertainment amenities of the place. The next group of transplants also is drawn to the city for its tourism, but, for them, the pull factors relate more directly to their dream of becoming an elemental part of the glitz, flash, and glamour encapsulated by the Strip. As Howard Schwartz, the insight-ful owner of Gambler's Book Shop, told me cynically: "Las Vegas attracts a strange breed of people who want to go out with a flash." His statement applies to both tourists and locals. Chuck Ballard, who came to town in the late 1960s on a basketball scholarship to UNLV, confirmed the implication. He said, people come here "to find fame and fortune. . . . It's like how so many people go to Hollywood thinking they will make it as an actor." Like Holly-wood, Vegas has its set of dreamers.

Most of the locals that fit into this group bring with them experience, tal-ent, and a will to use that talent to make it in the entertainment business. But, some, like Lorena, a Cuban American most recently from Florida, sim-ply have neon in their eyes. When I asked her how she came to live here, she answered excitedly: "Let me give you some history." Her brother-in-law planned a vacation in 2000 to Las Vegas and invited Lorena and her husband to come. They did the same the next year. She told her husband, "I want to live here." Four years later she did. She said, "I like the Strip. I like everything!" and then offered a specific reason. When she attended a show and saw Eliza-beth Taylor in the audience she became enthralled by the prospect of living permanently near the movie stars. She works successfully as a home-health nurse, undoubtedly hoping for another chance encounter with a star on the Las Vegas Strip.

Craig Brookings and Ben Murrell fall into the category of men with tal-ent who hope to survive in Vegas's green felt jungle. Both men came to play poker. Brookings migrated first to Northern Nevada from Texas in 1979 to work in a mine with his uncle. When he was without work the next year, he went south to Las Vegas to do what he loved. He survived only a couple of months with the money he took away from the tables and headed back to Texas. He returned in 1989 to try his luck again, this time getting a part-time job to support his passion. He eventually grew tired of the smoke in the casinos, but he still enjoys making an occasional bid in tournaments with low entry fees at smaller Las Vegas casinos. He makes a living today running his own small business.

Ben Murrell's story is similar, only Murrell worked full-time as a gambler for many years. In 1970, at age twenty-four, Murrell left his hometown in eastern Montana for Las Vegas. He had visited the city before that, but felt it was time to try his hand at professional gambling. He did well, too. From the way he spoke I could tell he knew that business like any corporate CEO knows theirs. For around thirteen years, Murrell made enough money to buy a customized van and new boat; to rent an apartment in Utah for two months one year so he could hit the ski slopes every day; and to take lavish vacations, including a two-month trip to Europe and an excursion to New Zealand. All of this on a poker player's "salary."

But, Murrell said sadly, times changed. When runaway inflation hit, everything cost more, but the poker player's wages in the midlevel tournaments (his preferred circuit) did not. He compared this change to working a job today at 1970s wages. He also cited the change in the gambling climate (the mobs had made flexible rules that benefited the gambler, which ended when corporations took over) as another reason he retired from the trade. Burned out, Ben stayed away from tournaments for years. He would like to get back into the game, but said in order to make a career out of it today, he would have to be in the high-stakes tournaments all the time and put in forty hours a week at the tables. Instead, he is reasonably content with recreational one-day tournaments and running his own struggling small business.

Some Vegas dreamers have the same goal as their Hollywood counterparts: success in entertainment. As *Las Vegas Sun* reporter Jerry Fink explained, "Las Vegas, self-proclaimed 'Entertainment Capital of the World,' draws them like a magnet, these wannabe headliners in search of a marquee." The total number of such people is uncertain, but probably exceeds two thousand a month. Jaki Baskow, a busy talent agency owner, has stated that hundreds come to her office every month in search of job placement.[4]

Fink has profiled a few among the masses of entertainers hoping to find themselves under the Vegas spotlight. Singer John Garafalo and comic James "Mr. Bigfoot" Scott are representative. Garafalo gave up a blue-collar electrician's job to pursue his lifelong passion for performing. He had kept his hobby alive part-time for twenty years, performing off and on since grade school. He moved to Las Vegas in February 2006, but knowing that success would not be easy, he also signed on with the electricians' union. Garafalo performs for free any chance he gets and plans to find a steady singing gig within two years that will bring as much income as he makes with the union. "I'd like to do upscale singing. I'd like to do some Broadway," he related. "I never really thought about being a headliner. My goal at this stage is maybe being an opening act for a headliner. That would be fine with me. Then I would have achieved what I came here to achieve."[5]

"Mr. Bigfoot" also dreams of a gig on the Strip. He came to Vegas from Cleveland in 2001 after a variety of occupations. For several years he looked for steady work, ending up in Vegas with some engagements in town and a steady set of gigs on the road. Having found moderate success, Mr. Bigfoot now wants more. He hopes to someday host his own late-night comedy show broadcast from the Strip. But, still, he doesn't see Vegas as a good starting point for dreamers like himself. Instead, he advises them to go to New York or LA first to gain experience. "If they don't have the skill, there's no place for them in Vegas. There aren't a lot of venues in Vegas for those just starting out. . . . The casinos are no place for beginners." Frank Leone, who leads the Musicians Union Local No. 369, agreed: "With the notable exception of Wayne Newton, this town does not create stars. . . . We get the stars once they have been created."[6]

Fulfilling dreams was a struggle for all the entertainers Fink profiled. The same was true for the gamblers I interviewed and even the lovers of the flash and glamour of the Strip. But, the *dreams,* and the chance to fulfill them in Vegas, are still alive.

THE ACCIDENTAL LOCAL

Vegas is a town of many myths: that casinos hide clocks and building exits to keep people playing; that a woman was once electrocuted while walking on the Strip; that oxygen is pumped into casinos to keep gamblers alert; that a complex system of tunnels runs beneath the Strip hotels. Some such legends have a foundation in reality; others do not. Here is one more. Recall from chapter 1 Thomas Hull, the man who built El Rancho Vegas, the first resort-casino on the now-famous boulevard. Hull was driving on Highway 91 just south of the city limits when his car broke down. His walk along the road inspired what would become his groundbreaking Vegas resort. So the story goes anyway. Hull's walk is one of the *untrue* yarns of old Vegas, but, on another level, the story of his arrival is representative of a different Vegas myth that I found to be quite true in reality: that of the accidental local.

Another Tom—last name Dennis—mimics Hull's story, but with a different ending. R. Marsh Starks profiled Dennis in the *Las Vegas Sun* newspaper: "Tom Dennis visited Las Vegas in 1981 and never left. After a long night of drinking and gambling, he found himself trapped in the Binion's Horseshoe parking garage with a dead car battery and eight dollars in his pocket. He lived in the car for fifteen days before finally driving it out." Dennis told Starks, "The last 25 years I've been here, I've worked some, a lot for cash." Now he collects fares at a parking lot just a few blocks from the one where he was stuck when he first arrived.[7] Dennis is a prime example of a large number of Las Vegans who, often as a result of a visit or short stay, find themselves here for years, decades, or a lifetime.

Twelve of the locals I interviewed shared their own versions of accidentally or unexpectedly ending up in Las Vegas. Some people stay after detours in their original plans. Ronald Shaw was on his way to Mexico to take a job as a police officer in the late 1970s when he stopped for a while in Las Vegas, never to leave. Joseph Mani told me it was not an accident but "destiny" that brought him to Las Vegas. Mani had lived in Washington DC and New York before returning home to India. After a divorce catalyzed plans for a life change and relocation, Mani intended to go to San Jose, California. On his way, Singapore Airlines ticket in hand, he noticed that the carrier also went to Las Vegas. He asked an attendant if he could change his destination. She said yes, and after some meditation, Mani determined to make the change. Part of the reason was the word "Nevada." In his native tongue of Malayalam, spoken in the Kerala region of southern India, "Nivada" means "you come." Since arriving, he has striven to build economic and tourism links between the two places, especially given Kerala's role as one of the larger tourist destinations in India.

Similarly, Beth Foster and her husband Daryl left their hometown for a new life in the American West. Drought and hail pushed the couple away from farm life in their native South Dakota. After unsuccessful attempts at establishing a home in Denver and then Albuquerque, they remembered hearing stories about making a lot of money in Las Vegas and reset their sights. "So we pulled into Las Vegas at around 3:30 in the afternoon [in August]," Beth related nostalgically. "It was 118 degrees that day, I remember. We had leather seats in that car, so we couldn't stand it in there. I didn't know if the person would mind, but I just sat down on the lawn of a house. I noticed a sign there that said 'Psychiatrist' and said to Daryl, 'we better go in there and have our heads examined.'" A passerby, who also happened to be from South Dakota, overheard the conversation and said to them, "It's not that bad. You'll be OK." And they were. The Fosters stayed in Las Vegas, enduring many more hot summers, watching their children and grandchildren make Las Vegas their home too.

Another refrain is common among this group of transplants: "I was only planning on X number of months and I've been here Y number of years." Jon Kubiak came in 1986 from Chicago intending to escape the cold winters. He told himself, "I'll only stay a few years." He has been here more than twenty now, and, unconvincingly, says he still has plans to leave soon. Dr. Mark Lewin graduated from chiropractic school and landed an internship in an existing practice in Las Vegas. "What was a six-month internship," he said, is now his career "eighteen years later."

A final set of stories are of newcomers who, like the tourist turned local, visited the city before moving here, but for reasons unconnected with tourism. Take, for example, Patricia Joseph, a friendly woman whose now-diluted New York accent sneaked out on occasion as she told her story. Joseph was a successful career woman who worked variously in fashion and makeup, publishing, and travel after she moved to Los Angeles in 1960. I was going to ask what brought her to Las Vegas, but, unprompted, she said, "I'll tell you why [I came]. When I was eight my mom taught me ballroom dancing. . . . It became my love. When I got to LA, I was told I needed to learn West Coast Swing." So she did. Many years later, she went on, "I came to a swing dance convention in Las Vegas." "Did you plan to stay?" I asked. "No intent to stay," Joseph replied. She never even considered a permanent move to Las Vegas. But a friend offered to show her around town on one visit. "And she showed me another life. This is where I was supposed to be. I got an apartment, sight unseen, and I moved here [in 1989]." She added that at the time she was interested in a career change anyway, so the unplanned move fit into her more general life path. Today, she still goes dancing two or three times a week.

An educator, former state legislator, and Clark County commissioner, Chris Giunchigliani also decided to move to Las Vegas after a visit. This Chicago native started her teaching career in the early 1980s in Shawnee Mission, Kansas, where she also tended bar for additional income. While in Kansas, she surveyed her life and knew, if she stayed where she was, she would eventually feel stuck, doing the same thing for the next twenty years. Around the same time she and some friends came to Vegas on a packaged vacation deal. She had been looking into a master's program and had taken a couple of classes at the University of Kansas, so she decided to detour off the Strip and check out the program at UNLV. Two weeks later, she moved to Las

Vegas, enrolled at the university, and got a job as the first female bartender at Jillian's in downtown Las Vegas, where she felt at home with the many native Chicagoans who frequented the bar. She related her own surprise at the decision to come to Las Vegas: "We just came to check the place out. I had no intention to move here."

Valerie Smith had a similar story. After attending a university in Chicago as well as Fisk and Vanderbilt universities in Nashville, Valerie began teaching school in Delaware. In 1979 she visited her brother, who was stationed at Nellis Air Force Base. Her brother tried to convince her to come to Southern Nevada, plugging the many opportunities for teachers in the growing city. She explored some of those, but, she said, "I went back to my job and didn't think too much about it." On another visit, she submitted several applications and was offered jobs the next day. She decided to accept.

Smith remembered the unique application process in Las Vegas: "I was used to being interviewed in front of people with suits, and they were wearing Bermuda shorts and sandals." She was also surprised at the speed. Teaching jobs could be found in a lot of places, but "there was no waiting time [in Las Vegas] like in other cities. I applied in California too, but was eleventh on the waiting list for art teachers. [Las Vegas] appeared to be a place with a lot of opportunity to get in on the ground floor of things. At that time, if you came to the city with any skills you could make it." Smith saw her expectations realized, working in teacher development and helping to determine the art curriculum to be embraced by the school district. She stayed at that job for nineteen years and now devotes time to her art gallery in North Las Vegas.

THE VEGAS BOTTOM-OUT

When the amazing growth of Las Vegas is discussed, speakers and writers often quote statistics that tell of six to seven thousand people per month coming to the city. Indeed, such statistics perpetuate two other myths: that Vegas is *the* high-growth town in the United States (ignoring other population explosions in the Sunbelt), and that everyone who comes stays.

I will discuss the myth and reality of these numbers and their implications later; I bring them up here to make a simple point: the growth statistics, in combination with the pervasive perception of opportunity and success waiting in Las Vegas for anyone who wants it, don't take into account newcomers who bust in this boomtown. The truth is, as Sinatra movingly sang, "this town" can both make you and break you down. By sharing some stories of people who want to leave and others who have ended up on the streets, my hope is to show this other, underdocumented aspect of transient Las Vegas. Reasons for leaving, of course, are as various as those for coming. Although I do not want to ignore stories of people who leave Las Vegas for "typical" reasons—job relocation or a change in personal or family life—my focus will be on people who tried their luck at Las Vegas and were dealt a bad hand.

The result takes several forms. Some people find they can't manage personal and family finances amid the gambling and entertainment temptations in Las Vegas and end up leaving town to avoid the struggle completely. Some stay, resorting to life in low-rent, dilapidated, and drug-infested neighborhoods in the shadow of Las Vegas Boulevard. Others end up on the streets, homeless and struggling to survive. Estimates of the homeless in Southern Nevada range from twelve to fourteen thousand people, numbers that contribute substantially to Nevada's rank as the state with the highest percentage of homelessness. Homeless advocates link these high numbers

The north end of the Strip at Sahara Avenue and Las Vegas Boulevard. In the foreground is Naked City, an area so named for its reputation in earlier times as the home of many topless dancers and showgirls who suntanned in the nude. Today Naked City is a neglected, low-rent district with a large drug trade. In the background the new Allure Las Vegas high-rise condos tower above the southeast corner of Naked City. Allure follows in a line of other similar complexes that have contributed to what has been termed the "Manhattanization of Las Vegas." Photo by author, August 2007.

to the city's culture of vice and addiction. Charles Desiderio of the local Salvation Army said: "When we look at Las Vegas . . . the streets are paved with (casino) chips. Everything is inexpensive, colorful, glamorous. Jobs are plentiful. In reality, they aren't as plentiful as people would like to believe."[8] This present-day phenomenon is an echo of the rush to work on Boulder Dam in the 1930s, when many more than could be employed on the project showed up and found themselves homeless in a much younger, much smaller Las Vegas.

Not surprisingly, gambling is the most prominent source of loss and despair for Las Vegans. Chuck Ballard, who compared Vegas's attraction to Hollywood dreams, extended his analogy: "It doesn't always work out that way. They come and get into gaming and they can't get away from it and end up losing it all. The most dangerous thing is a person who comes to gamble, and wins."

One homeless man, whom Kurt Borchard interviewed for his book *Word on the Street,* told of an addiction to gambling and a desire to get away from casinos. Said Philip: "I've got to stay out of the casinos, 'cause that's [a] weakness with me. . . . I'll sit down at the table and I'll stay there until I'm broke. That's one of the things that keeps me homeless, unfortunately."[9]

Although gambling is the most obvious malicious addiction available in Las Vegas, one vice often is accompanied by others. Jon Kubiak, whom I introduced earlier, addressed this compounding effect. In a discussion of what he liked and disliked about the city, Kubiak lamented the methamphetamine problem, which he has observed firsthand in the city's gay community and even among colleagues in his white-collar job working with the Clark County government. He said, "[The problem] is highest in the West, and, with the transient nature of the city and the party atmosphere already built in, I suppose it is more prevalent here." He added that the whole culture of drugs blends in with the atmosphere already here.

Jerry and Albert both told Borchard that drugs and alcohol were major reasons they ended up on the streets. Albert, for one, was living the Las Vegas dream. He had been married seventeen years and had a good job as a chef at a Strip hotel, but drugs and alcohol, he told Borchard, "[brought] me out here to this point where all I'm doing is living day to day. And I pray to God and I seek the Lord to try to change my life."[10]

Not everyone who struggles with gambling, drugs, or other addiction ends up homeless, but each continually feels the effect of the web of vice in Las Vegas. John L. Smith told me of the many people he has known who came to the city looking for a life of glitz and hustle, only to break down. Many of them work hard, often within the gambling industry, but eventually get to a point where that life begins to grind: "They get a job, live the life, and eventually bottom out."

Matthew O'Brien, in an evocative portrait of life in the storm drains of Las Vegas, provided story after story of Las Vegans who, for some reason or another, did not realize their hopes and ended up in underground "homes." David's story stands out. Coming to Las Vegas in the late 1990s after his mother died and with plans to put his training as a heavy-equipment operator to work, David decided instead to "just drown my sorrows in booze." He told O'Brien that, in just over a month, he spent a $43,000 inheritance from his mother on "video poker and table games." Eventually he ended up on the streets, and now makes his home in a tunnel underneath the famous sign on South Las Vegas Boulevard. His addictions to drugs, alcohol, and gambling keep him from a goal of getting back to Montana eventually. Now, he carries on an existence "just walking into casinos and finding credits left in slot machines."[11]

Credit hustling is a common source of income for homeless people. It involves identifying and playing slot machines that gamblers have left with money credits remaining on account. In a metaphoric nod to the mining heritage of Nevada, the technique is also known as "claiming" or "silvermining."[12] Another aspect of such "mining" is simply looking for change left in slot machine coin pans.[13] Such a practice is becoming outdated, since most of the machines in major casinos now provide payouts in the form of a ticket redeemable at one of the cashier cages on the casino floor. "Though illegal—according to state law all abandoned money becomes possession of the house—it's popular among local street people." David claimed to have found three hundred dollars in a machine at the Mandalay Bay on one occasion.[14]

O'Brien documented several other homeless men's stories of finding hundreds of dollars in unused slot machine credits. One might presume that homeless credit hustlers would take advantage of this to find a way out of their situation. But the occasional wins typically just serve gambling and/or drug addictions, thereby perpetuating the problem and further illustrating the notion that vice follows vice. O'Brien summed up the plight of many Las Vegas "busts" after an encounter with another credit hustler, Jim: "I realized he was just another person trying to make it in the world—just another Vegas gold miner who'd only found dust."[15]

Another group of locals, although not on the streets, also struggle to survive in an environment that has not been economically favorable to them. Jason and Elizabeth Butterworth, for example, came to Las Vegas almost ten years ago for a new start and to be closer to family but now jump from apartment to apartment with their three children, barely surviving on a meager income after bouts of bad luck, drugs, and unemployment.[16]

When I was a teenager I thought it curious that Las Vegas seemed to have more pawnshops than any other city I had visited. Pawnshops, of course, are notorious last resorts for someone to get a quick buck to feed an addiction. I have often wondered how much of that notoriety is myth, especially as it concerns locals. But I got a small taste of the punishment Las Vegas can dole out in an experience as I browsed at a used bookstore on East Charleston.

I drop by Academy Fine Books whenever I have the chance. I enjoy conversations with the owner, his collections of vintage Vegas gambling paraphernalia and local-interest books, and the nostalgic smell of old books, which reminds me of hours spent as a student in university libraries. As the owner and I conversed on this particular occasion, we were interrupted by a couple hoping to sell a used copy of one of the *Harry Potter* books. They assumed it would fetch a nice price, given its leather binding and a signature on the cover by J. K. Rowling. The bookstore owner looked it up, found that the signature was only a facsimile, and offered them twenty-five dollars in hopes that he could sell it for around seventy-five. The woman didn't like that, left the store, but came back minutes later asking for forty dollars. When the owner declined, she departed again, only to return once more to accept the twenty-five bucks. The owner said to the woman in a concerned yet cynical tone, "It's not that bad, is it?" When the couple left, the owner told me that he gets people like that all the time, people who come in wanting him to act as a pawn: "Give me sixty dollars for something and I'll give you seventy for it next week." I wrote in my field notes that day: "You wonder how desperate someone is. Maybe gambling? Maybe bad luck?"

Tina Lewis, the single mom who moved to Vegas from San Diego for a new start following a divorce, has an unusual perspective. Not only is she a Vegas transplant who *has* reaped the benefits of life in the city, but she also works as the director of an organization that assists low-income and homeless families to find accommodations that will keep the family together (most homeless shelters are separated by gender, forcing fathers to go one way and mother and children another) while they work toward permanent employment and homes. She told me of people she knows who have bottomed out in Las Vegas. The passion evident in her long, detailed stories, I think, illustrates the difficulty faced by many people who come to Las Vegas for its opportunity:

> What usually happens is people take their car, sell all their stuff, and take the couple of thousand dollars they have and say, "Let's go to Las Vegas, we'll get a hotel, then an apartment and we'll be fine." What really happens is they get here and get into a weekly motel. And a lot of them don't gamble at all. But, they don't get a job right away. And even though there are a lot of jobs, the majority of them are below $8.50 per hour, and sometimes they are unqualified for the higher paying jobs. . . . And they need a sheriff's card[17] for most of the service jobs, and other forms of ID that they wouldn't need otherwise. Most of the places do a credit check. . . . So, they need a sheriff's card, a credit check and fingerprints, and that's expensive. Then a lot of people can't pass the screening. They left tickets from where they were or they have warrants for their arrest and they came here for renewal. . . . Then they need to get the car registered and they run out of money and get stuck.

The weekly motels of which Lewis spoke are, like the pawnshops and casinos, a ubiquitous part of the Vegas landscape. Such establishments seem to have a more pronounced footprint in Las Vegas. For example, Siegel Suites, which operated fourteen Siegel Suites properties in 2012, exits only in Las Vegas, Mesquite, and Reno, while the other player in this industry, Budget Suites, had just three locations in Phoenix compared to the four in Las Vegas, which has less than half the population of its Arizona neighbor. While offering shelter to people who cannot pay a whole month's rent—with accompanying deposit, credit check, and utilities—such living arrangements can easily become a trap.

As my kids played alongside her grandchildren at a local park, Judy Haynes explained the perils of weekly motels. She herself had spent time in one. Haynes has had a trying experience in the city. She came to Las Vegas from New Mexico in the mid-1990s after a divorce from an abusive husband, but didn't find an immediate solution to her difficulties. She said that people often end up at the Budget Suites after hostile apartment evictions. In a tone that confirmed her hatred for such establishments, she said that drugs are rampant there. Her grandchildren's mother was one such victim, falling into trouble with methamphetamine abuse. This led to criminal activity and her current situation of serving time in a mental hospital instead of going to jail.

I met a handful of residents who had finally grown tired of whatever bad hand Vegas had dealt and were leaving, or getting ready to leave. Charles, an

The Las Vegas Strip is known for its flash and glamour. But just behind that facade is a two-mile stretch of road that does not fit this character. Zigzagging behind casino properties, the road is used for freight storage, backdoor casino entrances, employee parking lots, and apartment complexes. This view near Koval Lane and Flamingo Road shows the ubiquitous "weekly or monthly" rentals, as the for-rent sign indicates. Photo by author, June 2007.

African-American man from a small town in Texas, came to Las Vegas by way of Salt Lake City a year before I met him. His brother-in-law encouraged him to come, saying that he had an arrangement for making money. Charles and his wife came, but ended up earning only four dollars per hour in the scheme. He was headed back to Salt Lake that week. In a tone of conviction and detestation for his home of the last year, he told me: "I'm done here. Stick a fork in me." When I asked pointedly if gambling was the reason, his response was, "Everything. I knew we never should have come to Sin City." He explained that people he encountered would promise him one thing—such as his job—and that thing would never happen. Then he told me: "Everything will stick you in this desert."

Al Zanelli, the man who wanted to escape his stalking girlfriend, told me of his plans to leave. I asked what was keeping him here. He cited his vice and his punishment: "I want to go somewhere I can save some money and the machines won't take it all. You know. The bills are all paid, but you go to the bar and they're [i.e., the slot machines] right there. At the grocery store too."

Reasons for leaving the city don't always involve bad luck or economic struggle; sometimes Las Vegas just simply is not what a newcomer expects it to be. Many of the tens of thousands of retirees who have moved to town in the last decade eventually opt out. Some are not happy with the difficulties that accompany rampant growth. Others find that fixed incomes cannot keep up with the increasing costs of living. Still others miss family and friends they left behind and don't find the sense of community for which they hoped. Still other newcomers lament the city's still-lagging arts, culture, education, and health-care offerings. Such factors emphasize that departing Las Vegans are not only those who can't make it in this place of seeming endless opportunity. Eventually, other perceived difficulties in the city can drive out even successful locals, who, like Sacramento-native Luana Rae Bello, say: "I just have to leave. I'm outta here."[18]

Whatever the circumstances, stories of departed locals highlight the fact that Las Vegas is a land of comings *and* goings, a place of growth but not

The north-facing side of the famous sign. No silver dollars are shown here, only a partial outline of what the welcomed visitor (or local) came with. Photo by author, July 2005.

without an accompanying, albeit smaller, contraction. The backside of the sign I described at the beginning of this chapter is illustrative. The often-ignored, north-facing side of this marker bids the traveler a safe journey. More important, perhaps, is the absence of the silver dollars that provide a backdrop to the word WELCOME on the south-facing side. In relation to tourists and locals alike, the message seems to be: People come to Las Vegas, welcomed with their money and dreams, each of them ready for "fabulous" success, but many also leave, fewer coins to count and dreams broken.

A TRANSIENT CITY

Stories of going (as well as coming) point to transience as an important trait in Las Vegas's character and sense of place. According to Nevada State demographer Jeff Hardcastle, for every two people who move to Clark County in a given year, one leaves. More precisely, county-to-county migration data derived from address changes on yearly federal tax returns show that between 1990 and 2010, an average of 1.3 people move out for every two who move in. Such a pattern of transience takes concrete form in the experience of locals.

The local public radio station, for example, has trouble with fund-raising because "the audience is constantly moving [so that] many listeners are not around for the three to five years it will take them to donate. . . . Nevada Public Broadcasting has one of the lowest retention rates for first-year members of any public radio station in the country."[19] This phenomenon may be related to another aspect of local transience: longtime Las Vegans and newcomers who find a measure of success move around the valley with great frequency. In a 2009 survey, for example, UNLV sociologists found that 62 percent of Las Vegans moved to their current home from somewhere else in the valley.[20] Shari Nakae's experience is emblematic of this trend. I interviewed her in her fourth residence since moving to the city in the late 1970s to teach school; as the city grew, each time she moved to the new edge of town.

Schools and churches witness the constant change firsthand. My son's first-grade teacher invoked a common metaphor as she described the unique circumstances of teaching school in Las Vegas: "This year I've been lucky to only have two [new students] come in during the year, but last year it was like a revolving door with kids coming and going all the time. Some kids would leave and come back. . . . Phone numbers change and they don't tell what the new one is. . . . It makes it very difficult." And it's not just students that come and go in the school district. The Clark County School District hired three thousand new teachers for the 2007–2008 school year, 70 percent of whom were from outside the state. Of the new teachers, the district typically retains only about half past five years.[21]

Jeff Howell, senior pastor at a nondenominational congregation in the valley's northwest quarter, explained a similar pattern. He said that people move through his church like they move through the town. Nellis Air Force Base is close by, so changing assignments bring new families in and draw others out. Some members come to the church to recover or heal from wounds that Sin City has inflicted, but then leave after gaining spiritual and emotional stability. Dr. Ian Kaiser of a local Baptist church referred to the difficulty in developing leaders from within the congregation. "Since the community is so mobile," he explained, "the leaders move away. In the five years I've been here, we have trained three sets of leaders. They've either moved, didn't like the heat, or retired." Some of this transience stemmed from the church's location near the city center, he said. A number of members moved not out of town but to the suburbs and too far away from the church to commute each Sunday.

Transience explains not just the difficulties evident in the preceding examples, but the lack of a sense of community felt by many residents. Like other religious leaders, American Orthodox priest Father Kent Sharp sees frequent changes in parishioners attending services, but he also observed how the city's mobile nature affects his personal sense of community: "It is a very unfriendly city, and I've lived in New York and New Jersey. People there are much more friendly. I have neighbors [here] who, after six years, still haven't talked to me. I think a lot of it is the huge turnover." Mavie Roberts, similarly, explained that in the eight years she has lived in her house at Silverado Ranch and Las Vegas Boulevards (in the far southern part of the valley), only one neighbor has been there since she moved in. Other neighbors move in and out all the time.

One might assume that in a city that changes as often as Las Vegas, residents would be less likely to attach themselves to the place. For example, a bookstore clerk remarked that it is difficult to tie himself to any one image of the city because it changes so much. That assumption, I found, is only half correct. Even in a city where nearly everyone is from somewhere else, and where people come and go all the time, many people still find at least *some* attachment to place. In my experience, I found the line between the attached and the nonattached often follows the division between the longtime local or native and the relatively recent newcomer. The stories of Jimmy Del Toro and Annie Abreu illustrate my point.

Jimmy Del Toro was born and raised in New Mexico. He went to culinary school in Dallas before moving to Las Vegas in 2002. After finishing his education, a friend lured him to the city, promising plentiful jobs for a good chef. Jimmy came without a job, but soon was hired at Fiesta Henderson, a medium-size locals-oriented casino-hotel in Henderson. Soon, a friend helped him land a more prestigious position in the bakery of one of the Paris Las Vegas restaurants on the Strip. As we chatted atop the Circus Circus parking garage while awaiting the implosion of the Stardust, Del Toro explained his disdain for Las Vegas. Even though he loves his job, he hates the city. He and his wife decided to live in Henderson, because he "didn't want to be anywhere close to Las Vegas." When I asked him why, he said, "there's nothing to do here. All you can do is go bowling and watch movies." He cited museums, a zoo, and family fun parks as examples of what the city lacks.

When we met again several months later at a café, he said, "I still hate Las Vegas," listing the same reasons, but gave me a little more information. He detests the city's preoccupation with sex and sanctioned deviousness. Then he hit a chord regarding his attachment to the place, which came down to his relationship with other people. Even though he has a community of biker friends he rides with on weekends and holidays, he hasn't experienced the overall friendliness he enjoyed in other places. At work, he blamed that on greed. "Here, you're just a number," Del Toro complained, and he told me of a coworker whose sudden heart attack elicited little concern from colleagues. In more general terms, he blamed the lack of personal ties on the community's transience. He said, in a tone of surrender: "It's hard to make friends here, because as soon as you make friends, they leave." Even though Del Toro has been in Las Vegas longer than many people, and even though he has a good job and relative success in his career, he still gets lost in the constant comings and goings within the city. For him that leads to a lack of place attachment.

When we met at a coffee shop on West Sahara Avenue, Annie Abreu explained her relationship to Las Vegas. Abreu's father brought his family to the city on Halloween in 1953, when she was two years old. Her father was a dealer who had worked in a Cuban resort and became part of the migration from a former international gambling and resort destination to a growing new one. She said, with pride, that her father was the youngest person ever to become a casino manager in Las Vegas, and did so with only a sixth-grade education. Many threads of our conversation pointed to the notion that Las Vegas has afforded people genuine opportunities for success. In terms of careers, she explained, the city offers a lot. As she went through different stages of her life, Abreu was able to do what she wanted: In her twenties she liked to party and had plenty of options; in her thirties and forties she got serious about a career and became a successful consultant; now in her fifties, she wants more time at home and is able to do so by relying on the strong business she has built. She said, "It's easy living here. Whatever your lifestyle is, you can create a life here. If you're a really social person, there's a lot for you. If you're not, and you're more of a homebody, then that's fine too." In sum: "You can be somebody faster in Las Vegas than anywhere else."

Abreu called herself "a big cheerleader for the city," and said very little negative about it, quite the opposite of Del Toro. Such pride, such confidence in your town is something that, I think, comes with longevity and rootedness. Abreu put that attachment and belonging to hometown in good perspective, "This is what I know, and I wouldn't go anywhere else."

Still, the line between staying and going, attached and unattached, is not solely dependent on period of residence, nor is time of residence the only indicator of attachment. Although many natives, locals who came when they were children, and even some relatively new transplants consider Las Vegas their hometown, other longtime locals, even some who came early in their career and were successful in it, still are unwilling to take on the title of "local." Underscoring this pattern is the notation on the nametag of nearly all casino employees identifying the state or city they are from, which is most often something like Omaha or Columbus or Billings rather than Las Vegas. Tracy Snow, if you recall, still considers herself a Kansan even though she has made a semipermanent home in Las Vegas. Catholic priest Father George Toomey, who came to Las Vegas in 1982, agreed with Snow. He said, "Part of me feels that Las Vegas will never completely be home." He opined that a good way to tell where home really is would be to ask, "Where will your bones be buried?" Former North Las Vegas mayor Mike Montandon confirmed that many longtime Las Vegans consider somewhere else their home, and then made a strikingly telling comment that answered Father George's query: "Look at me. I'm the mayor and if I died tomorrow . . . they'd bury me in Phoenix."

At the same time, the magnetic draw of Las Vegas, a catalyst for its transience and something quite unique to the city, is another element of belonging in a place. I can think of nearly a dozen examples among my friends and interviewees who left Las Vegas for another life elsewhere and came back, many of them without plans to do so. In other words, many former Las Vegans come to realize how much they enjoyed their life in the city. Louise Fishman tried to move away twice but was drawn back to her childhood home. Aric Walker, a native of the city, said, "I've left to live other places, and I don't know, for some reason, after you've lived here you can't live anywhere else." Three peers from high school days, whom I interviewed individually, all left town for college (and in one case the military as well) and decided to come back when they finished their degree and/or military service. One of them said around 70 percent of his friends who had families here came back, adding that "the sense of community they built here was hard to turn down." My own parents were much the same: they moved to the Midwest hoping to get away from the hustle of the growing city, but were drawn back four years later to a job market that was much stronger than that in the Kansas City area.

The reasons for returning are varied, but typically boil down to opportunity, be it career or investment pursuits, love of the twenty-four-hour nature of the town, convenient access to entertainment, needs in the family, or simple attachment to the town they know the best. Trish Allison—who grew up in Las Vegas, moved away, and came back following a career change and a divorce—gave a vague yet encompassing description of this magnetism: "they keep coming back because there is something they want that they can only find here." In the end, such a draw must be an attachment to the place that, whether implicit or explicit, is pervasive and powerful. And, simply put, sometimes we have to leave a place to understand our love and connection to it.

The preceding discussion leads to a somewhat foggy, ambiguous understanding of attachment and belonging to Las Vegas. Based on my interviews, it is impossible to predict which locals will develop a sense of attachment. But, building on the notion that place and attachment to place are associated with experience, it makes sense that attachment *is* (and should be) an inherently vague concept since every person's experience in place is different from everyone else's.

The difficulty that many newcomers have in feeling attachment to the place, however, suggests we reexamine the claim that transience is a key part of the city's identity. Wallace Stegner once defined the entire West (including Las Vegas) as follows: "Look at any book that is western in its feel . . . and you will find that it is a book not about place but about motion, not about fulfillment but about desire. There is always a seeking, generally unsatisfied." Stegner's statement, and his elaboration on sense of place in the West elsewhere in his writings, makes one wonder if Las Vegas loses some of its sense of place simply because so much of it is not "in place." In other words, if sense of place implies rootedness, an established set of traditions, and something that "is made . . . only by slow accrual, like a coral reef," then how can Las Vegas qualify?[22] Considering how a sense of place is created, such a question becomes even more confounding. After all, extended time and experience in a location are key components of creating place. Several sense-of-place studies confirm such an assertion.[23]

Perhaps the contradiction between sense of place and rootedness is a question of semantics. As I discussed in the introduction, the term "sense of place" has many definitions and can be seen from many angles. If one defines it as rootedness, as Stegner did, then establishment and permanency is an important element of the term. If, however, one views the concept as an intimate connection to the character and personality of a locale, as I have done in this work, then rootedness based on long residence is not necessarily a part of the equation. The length of experience may not be as important in the creation of a sense of place as the experience itself. If that is so, we can use Stegner's characterization of the West to claim that the motion, movement, and mobility so prevalent in the region's history is, in fact, one element of its personality. Considering the stories I have shared in this chapter, I argue that transience, the sense of change and movement, and the resultant difficulty some locals have in attaching themselves to the place only adds to the city's character. Transience *is* a community virtue in this case, an essential component of the city's sense of place.

Many of the characterizations I've made in this chapter, which might be categorized as mundane or ordinary at first glance, become noteworthy because they are more pronounced in Las Vegas than in the country as a whole. Las Vegas doesn't hide its character, as a city of vice, a transient community, or a place of opportunity where some succeed and some fail. When I asked Tina Lewis if Las Vegas produced more broken people than other places, she said, "We certainly have a very fertile ground to grow in." She admitted that her hometown of San Diego and other places had plenty of vices, too—her parents, for example, admonished her to stay away from the navy seamen stationed in the area—but concluded: "Here it is just a lot easier to find it because it's right there on the Strip." Las Vegas is an exaggerated case of the draw that brought people to former mining boomtowns and that continues to draw transplants to new communities all over the world: a chance for a new beginning in a new place. Referring to the man or woman who is drawn to Vegas by the siren's call, John L. Smith observed: "It wasn't that the sirens sang. It was that they sang every day."

CHAPTER
FOUR

Getting Along with Growth

The June 20, 1955, issue of *Life* magazine tells an important story about Las Vegas. On its cover was an image of two African-American showgirls performing at the Moulin Rouge, the city's first officially integrated casino-resort. Although this was a significant event at a time when racism was rampant, more significant for today's Las Vegas is the caption just above the dancers: "Las Vegas—Is Boom Overextended?" The article inside depicted escalation in resort construction and questioned the sustainability of such rapid growth.[1]

More recent prognosticators have perpetuated the notion that Las Vegas could not possibly grow any more. In the wake of the attacks of September 11, 2001, for example, many skeptics and commentators assumed that the short-term lull in air travel and tourism traffic around the world would have a long-term impact on Las Vegas visitation. The proliferation of gambling across the country has produced a similar argument. As far back as the opening of the first casinos in Atlantic City, the argument was made that the closer legalized gambling comes to people around the United States and the world, the less "need" they will have to patronize casino-resorts on the Las Vegas Strip. Critics also speak out against Las Vegas from an environmental perspective. The editors of *Fast Company,* a business trade magazine, asserted that the desert metropolis is one of the world's "Too-Fast Cities," an "environmental pileup in the making." The editors further wondered: "Can the casinos find enough water to fill all those pools?" In response, Patricia Mulroy, general manager of the Southern Nevada Water Authority, said, "Twenty years ago we were answering the same question. . . . This is cyclical. Every two or three years, somebody says Las Vegas is going to run out of water."[2]

For decades, Vegas boosters, such as Mulroy, have countered the critics' attacks by invoking such failed predictions as they point to the city's continuing prosperity. They have a point. It is nearly impossible, for example, to read the 1955 prediction of "high-pitched optimism" without chuckling when one considers what the Strip has become since then. But, more than a tool for boosters or an example of historical hindsight, the failure of these prophecies is a good example of the city's character. Perpetual and spectacular growth lies close to the Las Vegas soul.

Rare indeed is the conversation about the city in which the topic of growth is not mentioned. The subject emerged nearly sixty different times when I talked with locals, and usually the interviewee addressed it spontaneously (i.e., without specific questions from me). Clark County Commissioner Chris Giunchigliani said the city's growing pains are the biggest challenges facing the community: "I found [this concern] going door do door. We asked open-ended questions: 'What is your number one concern? What is your number two concern?' And we would just let them talk. In most cases it was growth and traffic." Former city of Las Vegas mayor Oscar Goodman concurred when I asked him about challenges facing the city. He said, "I think it's the rapid growth. We have problems of sustainability. . . . It's the roads, the traffic, the air, the mental health, the schools."

Vegas, of course, shares the trait of rampant growth with Southern California, Phoenix, Denver, Salt Lake City, and other Sunbelt cities in the South and West. When I brought this up in my conversation with Mayor Goodman, he responded: "Yes, but [the problems of growth have] accelerated themselves here. It took a long time for Los Angeles to become congested. It happened overnight in Las Vegas." Since almost everything written about the city in the last few decades has addressed some aspect of that growth (further emphasizing it as an important part of the city's character) a rehash of those analyses and conclusions serves no purpose. Instead, my analysis focuses on the unique aspects of growth in Southern Nevada, through statistics and maps of growth, impressions of its impact from the local perspective, and illustrations of success and struggle that follow rapid expansion.

STATISTICS, MAPS, AND MEMORIES

In the boom years, six to eight thousand people moved to Las Vegas each month. So said the Nevada Department of Motor Vehicles, which records the number of new drivers' licenses within the metropolis. This statistic is impressive but somewhat misleading since it measures only the inflow of people into the valley and does not account for the large number of people moving out, such as the two-to-one move-in/move-out ratio I noted in the previous chapter. Still, possibly as a result of the move-out numbers, the simplistic driver's license statistics match up well with aggregate annual growth numbers from the state demographer and the US Census Bureau. For example, in 2005, the yearly increase in driver's license count stood at 83,510 (nearly 7,000 per month), and the total estimated population gain that year was 81,053.[3]

For a more general picture of population growth in the valley, consider again the growth trend in Clark County since its first participation in the census in 1910. Only 3,321 people lived in Southern Nevada at the earliest official count. By 1940 the number stood at only 16,414. After World War II, however, population exploded to 48,289 in 1950, 273,288 in 1970, and 741,459 in 1990. More rapid increase came during the 1990s as the resort boom on the Strip took hold, triggered by the opening of Steve Wynn's Mirage. A major milepost occurred in 1994, when the county topped one million people. By one local estimate that number doubled in the latter half of 2007, but the official census numbers and the Nevada State Demographer's office put the 2010 population at just under two million.[4] Such staggering growth in such a short amount of time makes Las Vegas the youngest big city in the country.

Las Vegas has remained near the top of the Census Bureau's fastest-growing-cities list for several decades. Between the 1970 and 1980 censuses, nearly 190,000 people moved to the area, which translates into a 69.5 percent growth rate and a ranking as the twenty-first fastest growing county of more than fifty thousand people. During the next decade Southern Nevada became the second fastest growing metropolitan area in the country as growth continued at a clip of 60.1 percent, with nearly 280,000 people moving to the region by 1990. At Census 2000, the county's population jumped by more than 630,000 people at an astonishing 85.5 percent growth rate, which kept the city at the number two slot just behind St. George, Utah. Ten years later, 575,504 people moved to the city, yielding a 41.8 percent growth rate and making Las Vegas the third fastest growing metropolitan area in the country.

Growth rate needs to be understood in order to more fully comprehend how it impacts a place. The rate is calculated by first subtracting population at the beginning of the period from population at the end of the period. That raw population change is then divided by the total population at the beginning of the period. The result is, of course, a percentage. This measure is helpful as we look at population change over time, particularly as an indicator of growth's impact. At the same time, a high growth rate does not necessarily translate into enormous raw population gain but rather population gain relative to the base population of the city or region.

We might understand this distinction by likening growth rate to filling a bucket with water. Let's say the bucket initially contains one gallon and we add another half gallon. That is a 50 percent growth rate, one that makes a noticeable impact in the bucket because of the size of the addition relative to the starting point. But, if the bucket contains five gallons of water and we add the same amount as before, a half gallon, the raw increase in water volume is the same, but the rate of increase is now only 10 percent. With a smaller growth rate also comes a relative decrease in impact. In other words, even though the same amount of water was added, the visible effects in the bucket are not as great as when we had a 50 percent growth rate. So, as the base population of a city—the population at the beginning of the period in question—grows by roughly the same number of people as it did the previous period, the rate, and the associated impact, decreases even though the raw increase did not.

Las Vegas is both a uniquely fast-growing city and, at the same time, just another growing city in the Southwest. A huge growth rate, like the one experienced between 1990 and 2000 is an important indicator because it has an enormous impact on the community. One way to think of this is that, for every ten people in the valley in 1990, more than eight additional people moved in. But, when we look at only the raw numbers, like the six to eight thousand people moving to the city each month, the increases are not as large as other growing cities. While Las Vegas added 630,000 in the 1990s, the much larger Phoenix gained more than a million new residents and Atlanta nearly 1.2 million. Las Vegas, therefore, is not alone in experiencing the problems associated with growth that I will address in the coming pages, but given the math of the growth rate, Mayor Goodman was correct to say that some of those problems are "accelerated" in Southern Nevada.

Another set of statistics illustrative of growth's impact in Las Vegas comes from the Clark County School District. CCSD has gone from the nation's fifteenth largest in 1990 to the fifth largest in 2010 when it operated 357 schools hosting 309,893 students. Enrollment has increased significantly nearly every year for the past two decades. More than ten thousand new students joined class rosters each year during the high-growth period between the 1993–94 school year, when the district enrolled 145,327 students, and 2005–6, when those numbers rose to 294,131. During this same period the district was forced to build more than one school each month to keep pace with the growth. Some elementary schools serve so many students that 82 of the 206 operated on a year-round schedule in 2007–8.[5] In the year-round school system, students are placed in one of five "tracks," each of which takes several alternating shorter-than-summer "breaks" throughout the year in order to accommodate more children at one school campus. In other words, only four-fifths of students at each year-round school are in attendance at any given time throughout the year. An abundance of "portable" classrooms, external annex structures on school grounds, is another indication of the district's struggle with growth.

Keeping pace with growth has led to another unique concern in the city. Given the city's relatively short history, Las Vegas community planners have difficulty finding a large enough body of social and political leaders for whom they can name all the new schools. As a result, the district has turned to naming schools after living people, some still in early stages of their careers. The junior high I attended, for example, was named for Kenny Guinn after he had served in local industry and as the district's superintendent, but long before he became Nevada's twenty-seventh governor. Such a pattern has led to deeper concerns about the ethics of naming schools after living people. Some observers point to a hundred nominations for four schools and accompanying lobbying efforts by interested parties and worrying signals that "the process has devolved into a popularity contest." Others fear that naming a school after a living person may force the school district, at some future time, to revoke the naming in the case of criminal or immoral action by the person after they have received the honor.[6]

One final intriguing statistic about the school district is the number of teacher vacancies the district had to fill annually during the boom years. In the July preceding the 2005–6 school year, around 500 teacher vacancies were reported. A year later the numbers were about the same, but increased to 565 leading up to the 2007–8 term and to more than 700 by the same time in summer 2008.[7]

Mapping growth gives an areal view of sprawl in the Las Vegas metropolis. Using a geographic information system (GIS), I traced, or digitized, the outline of the built-up portions of the valley on a series of historic maps and satellite images in order to show the expansion of Las Vegas's urban footprint over time. Note, in addition to the sheer scale of expansion in a relatively short period of time, the postwar expansion along Las Vegas Boulevard in the 1952 map, and the broad expansion to include the Summerlin (west side of the urban area), North Las Vegas, and Henderson suburbs following the major period of expansion in the 1990s.

In this different, more geographical measure of growth, Las Vegas again ranks near the top of the list. By one measure Nevada was the sixth most

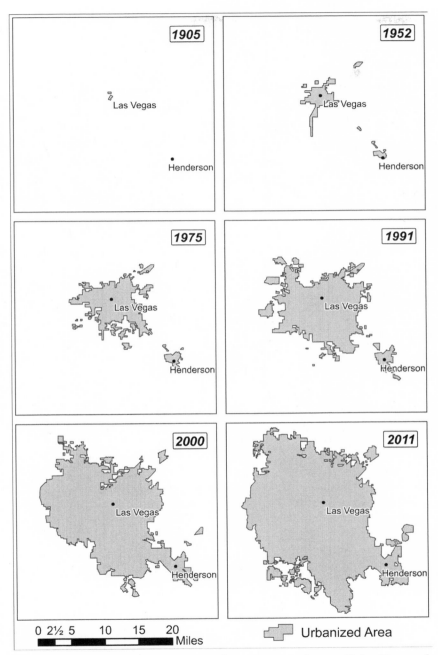

The Las Vegas urban area at various times between 1905 and 2011, with Las Vegas and Henderson city centers for visual reference. Maps by author.

These scenes show how the city's boundaries abut the mountain edges of the Las Vegas Valley. The top photograph shows a view of Henderson from the slope of Black Mountain. The bottom is a perspective from westbound Charleston Boulevard of a new residential complex that abuts Red Rock Canyon National Conservation Area. These and other projects highlight the dilemma of development versus preservation that cities across the country grapple with as they expand their borders. Photos by author, July and March 2007.

sprawling state and Clark County the second most sprawling urban county, behind Maricopa County, Arizona, where Phoenix is the county seat.[8] In almost any corner of the Las Vegas Valley, one finds neighborhood subdivisions with the curvilinear streets, culs-de-sac, and cookie-cutter houses characteristic of post–World War II suburban construction. This saturation of relatively newer neighborhoods is not surprising considering that in 1940 just over ten thousand people lived in the valley that now holds two million.

What follows in the wake of the population explosion and urban sprawl is an increase in land and infrastructure costs, and in the monetary and social expenses of an extended work commute. In response, leaders have pondered the residential planning of the future, asking: Do we let sprawl continue in the model of Los Angeles and Phoenix? Or, do we concentrate on high-density residential living? It appears that the reality will be a combination of both. High rises have sprung up both along the Strip—mainly for tourists and well-to-do outsiders to have a second home on the Boulevard—and downtown. Condominium complexes also have been constructed in the valley's urban fringes since around 2005. At the same time, developers of suburbs in these same fringes are shrinking the lot sizes of single-family homes, resulting in a higher density of housing per square mile. One real estate industry researcher pointedly summarized the close proximity of these newer single-family structures as follows: "Las Vegas is already one of the densest cities in the country. You can spit on the house next door."[9]

In 2006 and 2007 many people found salvation from skyrocketing home prices and dense housing by sacrificing a shorter commute for a less-costly home outside the valley. Older communities, like Mesquite eighty miles to the northeast, and Pahrump, forty miles to the west, have become bedroom communities. Logandale and Overton, two historically Mormon communities in Moapa Valley between Vegas and Mesquite also have boomed. Benny Gordon, a retired teacher living in Logandale told me: "This valley is changing. That is inevitable." He commented that something like 80 percent of the people living in Moapa Valley commute to Vegas. The evidence, he said, is the increased traffic on the main road through the valley.

Amid concerns about the availability of water, entrepreneurs keep planning the next exurban frontier. Northwest Arizona, north of Kingman and southeast of Hoover Dam, is one such site where the new communities of White Hills, Golden Valley, and Redlake are platted. A newly established Nevada town is Coyote Springs, where Reno businessman Harvey Whittemore wants to build 124,000 homes over 42,800 acres fifty miles north of Las Vegas. Commuters live as far away as St. George, Utah, nearly two hours (one way) from work in Las Vegas. Some observers insist that Southern Nevada will never sprawl as much as Los Angeles. But it is hard to deny the trajectory, as encapsulated in the comment of a local banking executive: "There's plenty of room in Pahrump to become another Victorville, California."[10]

A more subjective measure of growth in the valley lies in the memories and nostalgia of local residents. The sense of constant change is palpable.

A few examples of such stories include memories of a small town where nearly every Las Vegas student was able to walk to school, of how "there were 50,000 people here when we got here [in 1956]," of how "I could have bought acres of real estate for pennies compared to what it's worth today," and of the common practice of drag racing and cruising on Fremont Street in the 1940s and 1950s. And, as I mentioned in chapter 1, nearly every Las Vegan (newcomer and native) has their own story of when this or that road was dirt. The stories I will share here encapsulate the nostalgia that is so common in many conversations locals have about their place.

In a corollary to the stories of dirt roads, two interviewees pointed to the location at which we were meeting and said that it used to be bare desert. Eloise Freeman commented that only eighty thousand people lived in the valley when she moved here in 1958 and "there were only a handful of neighborhoods." She pointed around us to several areas that had been patches of empty land, including the spot where a coffee shop stands today at Sahara Avenue and Rancho Drive, only a little more than a mile from the Strip. Darrell Torrella, a career casino manager who came to the city in 1958 from Ogden, Utah, told me as we met in Summerlin about twelve miles from the Boulevard: "When I came here . . . you would need a horse or a helicopter to get to where we are sitting right now. . . . There's nowhere in the world that is as dynamic as Las Vegas." I had to chuckle at the image Torrella's comment evoked, realizing the relatively short time it took to bring about such a change.

In some cases, the city has grown so fast that newcomers cannot understand the historical background or significance of where they reside. At the other end of the valley from where Torrella and I met, Henderson sits as a good example of such a phenomenon. The suburban city's roots as an industrial town built around Basic Magnesium Inc. are not the most glamorous. Louise Fishman went so far as to call the town she was not proud to have grown up in "Hooterhell," a play on the "Hooterville" nickname the city held in its early decades. When newspaperman and entrepreneur Hank Greenspan's American Nevada Corporation began to develop the Green Valley master-planned community within Henderson city limits, the goal was an ambience independent of both "Sin City" *and* this blue-collar, industrial suburb.[11] American Nevada did such a good job of branding that people living in Green Valley today do not understand that they also live in Henderson.

Flint Salvador, a Henderson firefighter, explained the confusion in that part of the metro area: "We will sometimes get a call out to a home in Green

Valley and the people will say, 'We want someone from Green Valley Fire Department, not Henderson Fire Department. We live in Green Valley, not Henderson.' We tell them that Green Valley is part of Henderson and they don't believe us. So we end up getting a call back to the same place and have to tell them again that Green Valley *is* part of Henderson, but they still don't believe us." The Green Valley–Henderson confusion is further complicated as Henderson is changing again, its population becoming almost completely white-collar.

Ted Burke told me about his childhood in Las Vegas. A rare native born in 1944 in the valley, Burke's memories from half a century ago were lucid and sharp. He recalled delivering newspapers to Ed Von Tobel at his home at 105 N. 4th Street, which is in the heart of the casino district today. (Impressively, Ted recited several other Las Vegas founding fathers he delivered papers to, including their exact addresses.) My favorite among the many stories Burke told concerned a Boy Scout overnighter he attended with his troop. On their way out of town, one of the trucks got a flat and the spare was flat too. The leaders decided to leave the boys at that spot while they went back into town to fix the problem. It was so late by the time the trucks came back that the troop decided to stay and camp where they were. It was "way out of town," Ted said, and he described a huge mound of dirt on which the boys played "king of the hill." That mound of dirt was at Charleston and Jones Boulevards, a site only about five miles from downtown Las Vegas and many more than five additional miles from the edge of town today.

I have my own memories of change in the city, several of them involving the ever-evolving Strip landscape. As a teenager I watched as a high-rise addition to the Stardust was built in response to the boom on the Strip that followed Steve Wynn's Mirage opening. Sixteen years later, just after midnight on March 13, 2007, I stood atop the trunk of my car and saw firsthand (and in just a few seconds) the crumbling shell of that same building. A huge cloud of dust and debris expanded and moved toward the crowd, the image reminding me of a pyroclastic flow. After the dust cleared, I looked in the direction of the ruins while another view opened: Wynn Las Vegas, a new, lavish high-roller casino, in full view. The Stardust's replacement, Echelon Place, again will mimic Wynn's most recent transformation of the Strip to elegance and five-star hospitality. In a matter of minutes, in full Vegas fashion, I witnessed death and a harbinger of rebirth.

NEWNESS AND LOSS

The Stardust implosion held great significance for Las Vegas residents and tourists alike. The resort was one of the few left of a group built during the Strip boom of the 1940s and 1950s: the Dunes, Sands, Desert Inn, and others had already faced implosion. The Stardust's slightly longer run contributed to the nostalgia felt for it. It reminded patrons of the days when the mob ran things. It was a place where they could go and experience an old-style hospitality.

Las Vegas, of course, changes to please the hand that feeds it, whatever the consequences may be. My reason for sharing the Stardust experience is, in part, to show that whenever the city gains something new, marketable, profitable, and luxurious, it inevitably loses part of its historic character: it gives up the old and the nostalgic for the new and provocative.

This culture of newness experienced on the Strip bleeds into the local's realm as well, a direct function of the city's rapid growth. Each of the areas of expansion shown in the earlier map set obviously includes new homes, new streets, new businesses, new places for entertainment, and new casinos. Locals such as Russell Busch enjoy the freshness. After commenting on the latest locals casinos and their amenities, Busch told me: "That's another thing . . . everything is new here. People complain about that, saying there's not the tradition and culture of older stuff, but I don't think that's true. I like that they take this old building down to put up a new one. . . . I think hundred-year-old buildings are great, but for eighty years, it's just kind of crummy."

Near Oso Blanca Road and Gilcrease Avenue, facing south. At different stages of completion, this neighborhood in northwest Las Vegas is a metaphor for the region's ever-changing landscape: constantly growing, constantly morphing to new expectations of tourists and locals, and constantly looking to the future. Photo by author, June 2007.

When I asked Jonas and Cora Salvador what they saw as unique in the city, Jonas was quick to comment on the newness in the residential market. "Everything is new," he said. "There's not a lot of old stuff. In my business [as a salesman for kitchen and bathroom fixtures for high-end homes] you see a lot of new stuff. In other places there are a lot of remodels. Not here. There's very little of that." He and his wife, who live in a custom home in northeast Las Vegas, continued reminiscing. They talked about the neighborhood they both grew up in on Eastern Avenue and Bonanza Road and other "old" neighborhoods that used to be in the nice parts of town, like the Huntridge Addition at Maryland Parkway and Charleston Boulevard and Rancho Circle off of Rancho and Alta Drives.

Reading between the lines of the above two conversations, we can identify two unique aspects of Las Vegas's culture of newness. First, the neighborhoods the Salvadors mentioned are not "old" by the standards of other communities. The Huntridge neighborhood, for example, was built in the early 1940s in response to expanded activity at Nellis Air Force Base and Basic Magnesium Inc. Rancho Circle came along in the late 1940s, in an area that then was on the town's western edge. We see an exaggerated form of this relative "oldness" in the Stardust tower's demise following a mere sixteen-year existence. Busch's comment on "old" buildings also confirmed the Vegas definition of antiquity. Indeed, the general rule (and joke) in Las Vegas is that something is old if it was built prior to 1950.

The second aspect of the culture of newness evident in these conversations is a pervasive disregard for the (relatively) old landscape. Busch, a native, doesn't mind that old things are razed. Neither do most residents who move within the valley, like the Salvadors and other longtime Las Vegans. Jeremy Mont's comment sums it up. Mont came to town as a young child with his family from Biloxi, Mississippi. In his thirties now, Mont offered his obligatory comment on the increase in population and then stated with pride and zeal, "Las Vegas is not at all afraid to change. We blow [something] up and build something new in its place."

The Strip's blow-up-and-build-new culture is also readily visible in frequent newspaper articles that profile the loss of historical architecture in the city. Representative quotations show parallels with American consumerist culture. Here, though, rather than the typically "consumed" electronics or automobiles, the scale is exaggerated to buildings: "Las Vegas is widely known as a city without sentimentality, a place where the recent past is blown up and hauled away to make room for the next big thing." "Preservationists have wept openly as wrecking balls chipped away at Las Vegas history." "Disregard for Las Vegas' architectural past has broken many historians' hearts." "It could have ended up just another Las Vegas story, a memory faded from disuse, forgotten by the hordes of residents coming and going, then finally falling victim to the wrecking ball."[12]

The quotations indicate that the loss of the old in favor of the new is not acceptable to all Las Vegas residents. Preservationists, for example, have lamented the loss of several "railroad cottages" built near the historic depot between 1909 and 1911 to house workers. Other early homes have been razed or remodeled in favor of law and business offices. City leaders have pushed to establish historic status for such neighborhoods but meet opposition by landowners who believe restrictions on modifications to historic structures to be overly burdensome. In the case of the downtown home of Charles P. "Pop" Squires, a Las Vegas founding father and well-known newspaperman, the city council found that preserving the building by moving it to a permanent location would be too expensive. Instead it faced demolition and replacement by an office building for law offices and a private investigation firm. Kiel Ranch, which is listed on the National Register of Historic Places, also has met preservation challenges. This structure in the middle of a North Las Vegas neighborhood is one of the oldest in the state and the site of the infamous murder of Archibald Stewart in 1884. Still, the failure to amass the necessary funding has thwarted city leaders' efforts to save it. Nor has funding been successful to save the Huntridge Theater, a Streamline Moderne structure built in 1944 and also listed on the National Register. The owner wants to preserve the old movie house and concert venue in its original form, but has been unable to find a tenant and now insists he must find a more profitable alternative. He hopes to work with the Nevada Cultural Affairs Commission, which has pledged $1.6 million in restoration funds, to possibly convert the building into a restaurant or nightclub. Until then the historic structure sits, waiting for a possible rebirth.[13]

One of the best examples of failed preservation is Frazier Hall on the campus of the University of Nevada, Las Vegas. The first permanent building on the campus of what was then Nevada Southern University, this hall was named for Maude Frazier, longtime local educator, advocate for a state-supported institute of higher education in Las Vegas, and stateswoman. When Frazier Hall opened in 1957 as the classroom, library, laboratory, and administrative building for the new school, it was a "symbol of achievement for Southern Nevadans . . . the pride of Las Vegas." But, despite the best efforts of many people who tout the building's midcentury modern style, Frazier was razed in favor of a park that will serve as an attractive gateway for today's much larger university. In 1957, only five hundred students attended Nevada Southern. In 2008 most of the thirty thousand who attended UNLV had no idea of the building's historic importance to the campus and considered it nothing more than an "eyesore."[14] With so much of Las Vegas's population so new to the city, preserving historical landmarks becomes an almost impossible task.

Still, the trend in recent years appears to be toward more rather than less preservation. Residents and leaders increasingly want to separate the local culture of historical appreciation from the culture of the Strip, in hopes of using that history to bolster a deeper connection to place and sense of community. A vocal group of advocates, in fact, has been successful in preserving some of the city's natural, cultural, and architectural history. Red Rock Canyon, operating under the administration of the Bureau of Land Management as Nevada's first national conservation area, sits at the edge of the valley's urban development and is an important outdoor recreation site for locals. This park's sandstone formations and many ancient petroglyph etchings serve as a reminder of the region's natural and Native American history. Ironically, pieces of an earlier Strip landscape—including chunks from the old El Rancho and a demolished Caesars Palace parking structure—have been used to preserve and manage flood control along the Las Vegas Wash, which has helped to foster habitat for local wildlife.[15] Additionally, great effort has been taken by the Nevada Division of State Parks to preserve and showcase the Old Las Vegas Mormon Fort as the site of the first non-native settlement in the valley.

Redevelopment in downtown Las Vegas has prompted renewal efforts for several historic structures. In December 2006, a rare deviation from the typical course of events occurred for a now-dwarfed motel on the Las Vegas Strip. The architecturally unique lobby of the 1960s-era La Concha Motel was dismantled and moved to the Neon Museum's Neon Boneyard Park, an open-air showcase of some of the city's historic hotel and casino signage. The shell-shaped structure designed by Paul Revere Williams was reconstructed in 2008 and opened in fall 2012 as the museum's lobby. Former mayor Oscar Goodman has said, "Neon signs are to us what jazz is to New Orleans."[16] Just as the Crescent City, and others like it, preserve what best symbolizes the community, it is fitting that a preservation project to save uniquely "Vegas" things like neon and an old Strip motor hotel is one that has moved forward with such success.

Two other projects in the downtown area are indicative of this changing trend toward preservation. The historic post office that doubled as a federal courthouse was built on Stewart Avenue as a public works project to accompany the construction of Hoover Dam and later New Deal efforts in the valley. This building, also listed on the National Registry, has been renovated and restored to house "the Mob Museum," more officially known as the Las Vegas Museum of Organized Crime and Law Enforcement. It showcases the city's role in Prohibition, organized crime (including connections to other cities' mob activities), money laundering, and the Kefauver hearings on organized crime, some of which were actually held in the building. The project will contribute another element to the city's Cultural Corridor in which the Mormon Fort and Neon Museum also play a part.[17]

The Fifth Street School in downtown has been preserved and is a nice juxtaposition to the loss of Frazier Hall. This set of mission-style buildings was constructed in 1936 and originally housed the Las Vegas Grammar School. Millions of dollars have been poured into the project by the Las Vegas Redevelopment Agency in hopes of creating a "cultural oasis." Several organizations share space in the renovated buildings, including students at the Nevada School of the Arts, which relocated from a site in the city's suburbs. The city's Cultural Affairs Division along with a local office of the American Institute of Architects will also be part of the new campus. Finally, about thirty UNLV fine arts students will take classes at the site of the School of Architecture's Downtown Design Center. Mayor Oscar Goodman, a huge proponent of the project, has referred to the Fifth Street School as "a centerpiece of intellectual energy" for the city.[18]

KUDOS AND COMPLAINTS

Just as one Las Vegas resident may take the side of preservation while another advocates newness at the expense of history, the stories told by locals about other outcomes of growth fall along a similar spectrum of pros and cons. Accompanying sprawl, for example, are predictable complaints from rural and semirural dwellers around whom the new subdivisions are an unwelcome sight. Flint Salvador, the Henderson firefighter, lives in a northwestern neighborhood near Lone Mountain Road and Durango Drive still known for its horse culture. He spoke of riding horses and motorcycles there in his childhood, but now looks forward to the day when he can relocate to a more rural residence: "Vegas has gotten too big. When I was a kid, we had space, now we have too many neighbors. Not that the neighbors are bad, but the way I look at it—and you can write this down—when you can't go outside to get your mail in your underwear anymore, the place is too crowded."

Salvador is not alone. Charles Carpenter of Henderson lamented the loss of natural beauty on the valley's southern flanks in the face of new suburban construction. He grumbled: "For those of us who have enjoyed the sight of [Black Mountain] in Henderson, well, I can only say: It looks like this too shall pass! While driving on Horizon Ridge [Parkway], the natural sight of the mountains . . . will be with us no more. Yes, it looks like the earth movers are terracing the mountains to build houses. . . . I guess that it is just a matter of time until whatever natural scenery we have (had!) here will fall to the construction folks. What is next?"[19]

The Gilcrease family has watched sprawl envelop their orchard and ranching operation not far from Salvador's home. Originally some five hundred acres, the farm has been operated by the Gilcreases since 1920. As the city grew, the Gilcrease Orchard became a popular getaway for local families where they could see wildlife and pick fruits and vegetables in season. The farm has shrunk in the last several years, however, the family having sold chunks for schools and other suburban developments. Recently, things have taken a turn for the better. The Gilcrease family has been able to establish a foundation to preserve the remaining fifty-plus acres for future generations. Still, it is hard not to recognize the loss. Bill Gilcrease, who was a child when his family began cultivating the area, commented on the valley's sprawl: "the city came rolling over the hills and ran over us practically."[20]

Sprawl is only one of many negative outcomes that fast-growing cities experience. Traffic, which I will discuss in detail in the next chapter, is a huge issue, and water a close second. Darrell Torrella, the longtime casino industry insider, revealed this concern when he modified his initial statement that growth has "no end in sight" with the following caveat: "There's one end in sight and you know what that is: water. We're going to run out of water. We had an old saying in the gaming business: 'All you need in the desert is water and overtime.'"

For Aric Walker and others, air quality is another major concern. He told me: "It was a lot clearer when I was growing up. We'd get that 'Henderson cloud' [from industrial activity], but it seemed like there was always a breeze blowing it out . . . that's the biggest problem." Speaking of the burgeoning school system, retired teacher and mother of two Shari Nakae complained of the district's difficulties and faults, saying it has just grown "too big, too fast." Affordable and available healthcare is another source of angst for Las Vegans. Bonnie and Orlando Pratt, for example, complained about how long it takes to get in to see a doctor. In addition, living in the booming city became generally less affordable than it once was. A cost of living index from the American Chamber of Commerce Research Association in 2008 put Las Vegas at nearly 10 percent higher than the national average.[21]

City services and infrastructure share in the growing pains. When I attended the Metropolitan Police Department's Citizens' Academy, one officer explained that, when new neighborhoods are built, a certain amount of money is set aside for fire stations as a requirement for insurance purposes. Police get only a minimal amount, however, and as a result, the department has not kept up with growth well. Depicting the struggle Metro faces, one advertisement in a public awareness/officer recruitment campaign in 2007 proclaimed: "The city's growing fast. We're helping to see it's brought up the right way." The criminal justice system is also struggling. The Clark County Regional Justice Center ran out of space only ten months after it opened in early 2006, and existing jails often lack the required number of beds for inmates. The Las Vegas–Clark County Library District, like the police department, also has had to sacrifice and creatively manage its budget, especially after a library construction bond measure failed in 2003.[22] For some locals, the city has just become too big. Interviewees complained of a constantly expanding road system. Catholic priest Father George Toomey told me: "It is strange to live in a city where you don't know all the streets." You can drive to a part of the valley and not even recognize it, he continued. "It is a little unnerving." Shawn Newman explained how she would often get lost because of misinformation from an online mapping service: "MapQuest can get you all messed up . . . the roads not ending up where you thought they would."

Even as Las Vegas reeled from a global recession that brought the boom times to a halt, the city's residents continued to feel growth pains. It is true that many complaints from 2007 no longer had grounding in 2010 or 2011. In fact, homes became more affordable as prices plummeted, some infrastructure improvements may actually catch up with the resident population, and the water crisis is no longer as imminent. Yet, the rush to build homes triggered by runaway population growth has given way to a landscape of foreclosure and neglect. This has had no small impact on neighborhoods and residents that hoped to be part of an idealistic, master-planned community.

Nasser Daneshvary, who directs the Lied Institute for Real Estate Studies at UNLV, summarized the change: "We were the hottest market in the nation in terms of . . . how fast it went up. . . . And, of course, when something goes up, it comes down hard, too."[23] One colleague noted how his sister's neighborhood in Las Vegas seemed like a ghost town after so many abandoned their foreclosed homes. According to one census estimate, more than 167,000 homes were empty in Nevada in 2010. Of course, emptiness isn't the only result of such a crisis: neglect by former owners, or current ones who don't feel the motivation to take care of their properties, has generated a slew of code violation complaints and has given way to increased crime and drug prevalence in affected neighborhoods.[24] Again, the words of Daneshvary: "Foreclosed homes are toxic neighbors."[25]

Each of the issues I have cited here detracts from the overall quality of life in the Las Vegas Valley, a general topic that has garnered much attention in local media and research outlets in recent years. Newspapers report and opine regularly on such concerns, typically pointing to publications like the Annie E. Casey Foundation's Kids Count data book or Bert Sperling and Peter Sander's *Cities Ranked & Rated*. Letters to the editor often air the complaints of vocal residents, like Kathy Williams, who wrote: "Las Vegas continues to

"Hot potato homes" have become prevalent as the national economic recession has brought Las Vegas a glut of empty new homes and a decreasing population growth rate. Perhaps this developer of a small subdivision near Grand Canyon Drive and Patrick Lane foresaw the recession in August 2007 when I took this photograph.

be a great destination for vacations and conventions. . . . However, unfortunately, it is not such a great place to live." Williams pointed to many of the concerns I've identified and then concluded: "Many of us would leave if our financial situations would allow it. In the meantime, can't we find some ways to save this city and improve the quality of life for everyone?"[26]

Community and university researchers have weighed in on how deeply growth impacts the city's life and community. In 1998 Ronald Smith, a UNLV sociologist, edited the *Las Vegas Metropolitan Area Project,* with essays about several quality-of-life issues. A 1999 master's thesis by UNLV economics student Terri Hicks used a valley-wide survey to find that traffic, home ownership, and air pollution most prominently impact residents' satisfaction with living in the area. In October 2005, at the height of the boom, the UNLV historian Hal Rothman (whose writings often addressed the city's growth) penned a series of columns for the *Las Vegas Sun* in which he highlighted many of the quality-of-life issues increasingly influenced by the city's explosive growth. A 2007 study by United Way of Southern Nevada identified the most pressing concerns in the area as health care, education, affordable housing, and financial stability. Hoping to effect change, the Nevada Chapter of the American Institute of Architects published *Blueprint for Nevada* in 2007, with participants in Southern Nevada making recommendations regarding resources, economic development, arts and culture, education, and transportation. Similarly, UNLV sociologists published the *Las Vegas Metropolitan Area Social Survey: 2010 Highlights,* based on surveys and focus groups to identify local residents' views of urban sustainability from environmental, sense of community, and economic perspectives.[27]

The abundance of complaints by locals and the conclusions of research groups highlight the powerful role of growth in the city's personality. Yet, even though the negativities appear to be *the* overwhelming part of that personality, they are not the only result of growth. With the bad also comes the good, and with the struggles come benefits. Many locals see it this way, as I found often in my interviews.

A few optimists, like JR Henson, mentioned only the benefits of growth. As we sat outside a coffee shop on a pleasant, sunny February afternoon, he told me: "The great thing about this town is that nobody gave a damn what your pedigree was. They care who you are and what you can do right now. In other places, people would give me a bad time about Vietnam, but here nobody cared. They cared that you did it, but they take you at face value. Anyone can get a job in this town if they want to work. . . . The other thing is that it is a town of second, third, and fourth chances. You can screw up and leave and come back and make it again. . . . It's an expanding market."

Similarly, Leonard Lewis couldn't say anything negative about the city, even after significant thought on the matter. When I asked the forty-year resident and successful entertainment entrepreneur to describe living in the city, he responded: "It's exciting. It's energetic and futuristic. No matter what you see today, it will be better tomorrow. It's not that bigger is always better, but here it's worked out that bigger *is* better. . . . I don't know how to express it, but [he made a gesture of an explosion in his hands] it's energetic!" When I asked next, "What is your favorite part about living in the city?" Leonard said it was the opportunity. "No matter what you do, if you do it well, you'll be rewarded." When I shifted the topic to his least favorite part about living here, he responded, "What *don't* I like about it?" as if he didn't know the answer. "I can't say I don't like the growth, because that's what it's all about. . . . It has jobs and opportunity." Pausing for a moment, he said he'd come back with a better answer. Then he added, "If there was something I didn't like about it, [the city] wouldn't be what it is." If you say it's too big, he continued, then it doesn't have what you like about it.

Not everyone was as overtly optimistic as Leonard Lewis. Most interviewees, when asked about least favorite aspects, could name at least one thing. Often that thing was related to growth; and, more often than not, interviewees who addressed the growth issue gave me opinions both pro and con.

Two weeks after I spoke with Flint Salvador, who lamented the growing density of homes around his once semirural residence, I interviewed his uncle, Shirley Foster, in his nearby ranch-style home decorated in rodeo gear, which hinted at why Shirley (and Flint) settled in this area. Foster had views similar to his nephew. Suburban construction, he said, "is closing in fast on me." At the same time, he noted several positive aspects of growth. "There is always opportunity here because of the growth. Do you know that in 1956 there were four thousand people per month moving to Las Vegas and it hasn't stopped? Some months, seven thousand people move here now. The opportunities for people are unlimited." Then he noted both a good and a bad quality: "You know I like the convenience of Las Vegas. You can virtually buy anything you want. . . . The downside is traffic. It has grown so fast that they can't keep enough asphalt on the road."

Jake Glennon's experience has shaped a similar attitude about growth, but from a business perspective. Glennon owns a successful company that offers tours of Vegas and outlying areas such as Hoover Dam and the

Grand Canyon. When asked about his least favorite part of the city, he said: "I come to work with a smile on my face every day, and usually I go home with a smile." He pointed out that he doesn't have a lot to complain about. Las Vegas has been good for his business and he has done well. However, he recognizes some functional difficulties that come with an expanding business: "But, with the growth, it becomes a numbers game. You get so busy. You think, how can I put ten pounds in a five-pound bag? How can I fit this number of people on a helicopter?" Glennon said that his business has always struggled to keep up with demand, so much so that he has to turn people away. For most business owners, such a problem wouldn't be a problem at all, but to Glennon, his business has lost something he values. "Sometimes I find myself yearning for some of the old days. . . . It used to be that I knew everyone that worked for me. I knew their spouse's name. It's not that way anymore."

I witnessed such a back-and-forth in many conversations, and in almost every one the positive attitude toward growth focused on the opportunity (usually economic) that growth provides. The negatives were more varied but focused on quality of life. In terms of politics, the complaints often were leveled at local politicians' inexperience and what one interviewee called "myopia" and a "failure to understand how to deal with growth." A constant influx of people, however, can bring in new perspectives to help solve the problems. Nevada's large population increase, most of which has occurred in Clark County, gave the state a third seat in the US House of Representatives after the 2000 Census, and a fourth seat in 2010.[28] In addition, an influx of people from elsewhere also brings new cultural opportunities and ideals associated with an increasingly diverse population, an enriching aspect of the city that many locals appreciate.

The duality of the Las Vegas boom has been recognized by Las Vegans since the days of Hoover Dam's construction. The quiet of the small town was gone by the early 1930s, with old homes destroyed to make way for projects necessary for the "future metropolis." At the same time, newspapermen recommended thoughtfulness in the changes allowed by the town's residents. Pop Squires of the *Las Vegas Age* recommended: "In our zeal to keep pace with the growth that is coming, we should hesitate before we destroy those things belonging to our infancy, which add beauty and individuality to our city." Two months following such comments, Squires added: "if you wish to preserve the memory of Las Vegas of the past 24 years you better take your snapshots now."[29]

The dilemma of "old and historic" versus "new and flashy" highlights the ambivalence regarding growth in this city and, by extension, other places like it. The stories I have shared illustrate this dilemma. In fact, of the forty-four interviewees who mentioned impressions of growth in our discussion, twenty named *both* positives and negatives, eight gave only positive outcomes or impressions, and sixteen remarked only on the negative aspects.

Survey responses from the *2007 Las Vegas Perspective* give that same impression. The *Perspective* reported that the majority of locals surveyed were concerned about the valley's growth rate (83.2 percent), traffic congestion (72 percent), air pollution (87.5 percent), education (82.8 percent), and water availability (90.9 percent). At the same time, people recognized the positive outcomes of growth: 52.1 percent of survey respondents agreed that growth had been good for the region, and 44.9 percent that growth had been good for the respondent personally.[30]

Even though a "take the good with the bad" attitude prevails in both my informal survey and the *Perspective* results, the higher number and greater intensity of negative comments is significant. While Las Vegans typically recognize the double-edged sword of growth, they also recognize that one edge is sharper, resulting in what we might call an imbalanced ambivalence.

THE HOUSE OF CARDS

Regardless of how Las Vegans feel about the city's expansion, and despite recent changes in the city's growth trajectory, the fact remains that perpetual change is a big part of what makes Las Vegas distinctive on the stage of American culture. It is, in fact, common to hear conversation among locals about the city's persistent reliance on growth, almost as if it was an addictive substance Las Vegas must have to viably exist.

But such a comparison prompts questions about the end game of that addiction. Will the city, like a drug addict, hit bottom in the form of a total collapse? Is the growth environmentally, economically, or socially sustainable? Author and social commentator Mike Davis wrote a 1995 *Sierra* magazine article entitled "House of Cards." In it he criticized Las Vegas's "hypergrowth" and its impact on water and energy resources and escalating social problems, many of which I have mentioned here.[31] His use of the house-of-cards metaphor, however clichéd, provokes interesting questions regarding the Vegas obsession with growth and the ever-present potential for failure.

First, perhaps, is the question of how city residents and leaders should cope with the problems of growth, particularly those related to fostering a high quality of life. The answer is complex and, of course, is something Las Vegans will grapple with for years to come. I want to present some of the thoughts circulating within the community that may lead to a solution. On one end of the spectrum stand the "let growth continue but make growth pay for growth" advocates. This camp wants higher taxes for the general public, perhaps an individual or corporate income tax (currently absent in Nevada) or taxes and fees for real estate developers or an increase in the gaming tax (since the growth in tourism feeds so much of the growth in the valley). On the other end of the spectrum are "stop growth" proponents, some of whom have pushed for a moratorium such as a Portland-style "ring around the valley," which was proposed in the Nevada legislature in the late 1990s by then state senator Dina Titus, but eventually failed. One local resident had his own version of this solution printed on a bumper sticker: "Save Las Vegas. When you leave, take someone with you."[32]

Other solutions lie somewhere in the middle. A common idea involves technology. Just as air conditioning and giant "straws" that transport water from the depths of Lake Mead have kept the city habitable in a physical sense, ingenuity, technology, and the associated capital may continue to provide transportation for masses of people from distant suburban communities, reduce the harmful toxins automobiles expel into the air, and accommodate more students or patients with limited human and physical resources.

Recognizing that technology is good only if it remains economically and environmentally viable while at the same time subscribing to the notion that Las Vegas *needs* growth to remain successful, the next natural middle ground is sustainable or "smart growth," a part of the new urbanism movement. Smart-growth advocates push for planning that promotes environmentally sustainable, walkable, and mixed-use communities. This was one recommendation, for example, in *Blueprint for Nevada*. Other organizations and developers are embracing creative alternatives. The Regional Transportation Commission (RTC) of Southern Nevada, for example, initiated a billboard campaign to encourage commuters to use alternative means of transportation, and developers in new subdivisions like Inspirada in the southwestern Las Vegas Valley are building higher-density, walkable neighborhoods.[33]

Although no simple treatment for the Las Vegas growth addiction likely exists, city leaders and residents must grapple with the consequences of runaway growth and plan for the future. By implementing one alternative, or a combination of several, they may be able to squelch some of the negative aspects of growth while still reaping its rewards. The philosophy of Oscar

Goodman as stated in a 2007 interview illustrates the "Vegas can-do atti-tude": "I think the worst thing in the world would be for the city to artificially impose a growth measure to slow us down. Water is an issue. Traffic is an issue. Air quality is an issue, the homeless is an issue. . . . It's not a situation where you can take time off, and say it's going to take care of itself because none of it does. . . . It's just a question of staying with it. . . . That's the way I look at it. Nothing around here is insurmountable."[34]

Of course, the answers to the first question about how to cope, and even the debate about such answers, hinges upon a second question regarding the Las Vegas house of cards: How long can it last? Most observers have pointed to the city's past success and concluded that it will go on in perpetuity. But, even as I have been writing this book, events have caused me and many oth-ers to pause at such optimism: population actually declined in 2009; home foreclosures have risen; the casino gambling take has waned and with it state tax revenues; tens of thousands of jobs, many of them in construction, have been lost, yielding one of the highest unemployment rates among US cit-ies; and, as a sure bellwether of economic conditions in Las Vegas, manag-ers at the downtown Golden Gate hotel and casino have raised the price of the shrimp cocktail, a favorite for locals and tourists at less than a dollar for nearly fifty years, to $1.99.[35]

Still, many locals, including much of the media, the Las Vegas Convention and Visitors Authority (LVCVA), and other analysts and experts, see the city bounding out of the current economic difficulties and into a bright future. Such optimism is based largely on the tourism market, the main reason the city has successfully weathered recessions of the past. In February 2008 sev-eral major local papers printed stories about "The Boulevard Formula," which is that "Strip resort-building [is] a tried-and-true fix to Southern Nevada's periodic economic doldrums." The LVCVA, for example, projected that more than 41,000 new hotel rooms would be added to the Vegas inventory by 2012 as new hotel-casino-condo complexes continued to alter the city's skyline. Each new room, according to Jeff Hardcastle, Nevada state demographer, yields an estimated 1.5 jobs at the resort and another "indirect job" in the community (teachers, retail clerks, doctors, etc.).[36] Optimists claimed that would produce more than a hundred thousand new jobs, a boom that could bring in nearly a quarter of a million new residents to the valley, boost a fail-ing housing market, and save the Las Vegas economy.[37]

By summer 2008, however, even the Strip's ability to deliver the region from the recession seemed in question. Like a bomb, news on August 1 that Boyd

The shell of the suspended Echelon Place project on the former site of the Stardust. Photo by author, January 2011.

Gaming Corporation would be suspending construction on the almost $5 billion Echelon Place after a little more than a year in construction shocked people's confidence. The move, although initially good for Boyd Gaming's stock prices, brought an immediate loss of construction jobs and a likely slowing in projected job growth overall. In addition, the suspension probably will delay a rebound in the local housing market and check optimism about a quick turnaround. Indeed, the nationwide financial crunch halted other Strip projects. The Plaza Las Vegas owners announced two weeks later that its $5 billion project on the former site of New Frontier would not go forward until credit markets stabilized. Across the Strip from Echelon, the mammoth Fontainebleau, a nearly completed mixed hotel/resort/condo project, sits empty, its owners waiting for the right time to complete it or sell it off. And the Sahara, one of the Rat Pack hangouts built during the 1950s boom, closed its doors in March 2011 and also awaits an uncertain future.[38] Besides being a portent of what was to follow, the greatest impact of the Echelon delay, however, was to show in poignant fashion that Las Vegas truly is vulnerable to an economic downturn.

Given the vacillating confidence in Las Vegas's future as a growing, thriving metropolis, it's easy to wonder if the growth I have identified as central to the city's personality is not somewhat misplaced, especially following the so-called Great Recession. Such questioning is an opportunity to see the city's face from another angle. Las Vegas is still a growing city. It has shown its resiliency in the past. Although I would not go so far as to predict an untroubled future, neither will I ignore history and the ability of Las Vegas and its residents—using tourism as a bastion as it has since the days of Hoover Dam—to weather a storm. Many critics, like those who penned the June 20, 1955, issue of *Life* magazine, have bet against Las Vegas in the past and lost. So, regardless of how gloomy the situation looks, or how long recovery takes, I have to believe the city will find a way. The house, after all, always wins in the end.

Stuck in the Fast Lane

State Route 160 in the southwestern Las Vegas Valley is one of the most dangerous roads in the city. When Maria Guadalupe Martinez's car veered into oncoming traffic on the two-lane road in 2006 near the intersection at Arville Street and Martinez was killed in a head-on collision, it marked the eighteenth fatality on that highway in less than eight months. This number got the attention of state and local officials. The Nevada Highway Patrol allotted troopers to the busiest stretch of the thoroughfare, also known as Blue Diamond Road, and they established a "zero tolerance" initiative to ticket speeders and reckless drivers, aimed at putting a stop to what NHP spokesperson Kevin Honea described as "our deadliest beat for the last three or four years, at least."[1]

This action was only the beginning. A few days later, the Nevada Department of Transportation lowered Blue Diamond Road's speed limit from sixty-five miles an hour to forty-five miles an hour. A week later authorities announced a mandate for drivers on the road to use daytime headlights as well as plans to install "rumble strips" along the road's shoulders to alert wavering drivers, repaint shoulder and median lines, expand left-turn lanes at the busy Rainbow Boulevard intersection, and install traffic signals at Durango Drive. Suzan Hudson, a resident living in the area, took things further still. She placed twenty-one white crosses at the roadside locations of recent fatalities, hoping to remind political leaders and motorists "what's inevitable when an unforgiving road is packed with impatient drivers from new homes in what was once empty desert."[2]

The deadly Blue Diamond Road also sparked a heated conversation in the pages of the *Las Vegas Review-Journal*. Decision makers defended their Band-Aid measures (which were meant to prevent more fatalities until the road could be widened from two to eight lanes, which was completed in late 2009), while critics questioned the effectiveness of the proposed actions. Letter writers vented by cursing local planners for not doing something sooner to make the road safer or blaming the fatalities on bad driving habits. One vehement resident recommended that officers be equipped with bazookas:

"These speeding SOBs don't care about your life. Why should we care about theirs? All it would take is one shot, and problem solved."[3]

ROAD CONDITIONS

The story of Blue Diamond Road epitomizes the transportation struggles and attitudes that are part of the Las Vegas character. The confluence of construction delays, heavy traffic, heated tempers, and staunch opinions in that story underscores the experience of locals all across the valley as they encounter congestion and bad driving in their comings and goings. The magnitude of this experience is confirmed by the frequency with which the topic surfaced in my own research and conversations with locals.

Traffic is a major part of all local dialogue, rivaled only by the obviously interrelated issue of growth. The *Las Vegas Review-Journal,* for example, has its own version of the *Los Angeles Times* Road Sage and Bottleneck Blog columns. The Road Warrior is a twice-weekly column dedicated to information about traffic and transportation, updates on road construction, and answers to questions from readers. Omar Sofradzija, who formerly wrote the feature, described it as follows: "You have thoughts. I have space. Put the two together and you get a sense of what Las Vegas Valley roads are looking like . . . for the average schmoe on the go."[4]

Readers enrich (and sometimes inflame) the omnipresent conversation through letters to newspaper editors. Along with eleven letters I clipped about Blue Diamond Road, I found an additional forty-eight between summer 2005 and summer 2008 that addressed broader transportation issues. That number is certainly lower than what actually appeared in print (I was a selective clipper), and it does not include the scores of letters the Road Warrior receives and publishes each year in that column.

Transportation concerns were equally prominent in my interviews. Of the fifty-four people I asked, "What is your least favorite part of living in Las Vegas?" "traffic" was the most common response at nineteen times. Incidentally, the second least favorite, mentioned twelve times, was growth and its problems, which cannot be separated from concerns about the city's roads. Moreover, the traffic complaints came up in the natural course of many other conversations. Its power and influence, in fact, is difficult to overstate.

Concerns specifically related to poor driving came to the forefront of public discourse in 2005 when the Clark County Commission launched an initiative to raise awareness about the subject. Hoping to curb distracted driving, red-light running, and driving under the influence in particular, the commission sponsored televised public service testimonials from residents who received citations for their violations and recognized a need for a change. Additional campaigns included ads on billboards and bus panels, plus bumper magnets with the slogan: "Bad Driving... What's Your Excuse?"[5]

Rapid changes in the valley's road network further illustrate the significance of transportation issues in the city. While nearly every native *and* newcomer will tell the story of when this or that road was still dirt, other locals grumble about the difficulty of finding their way for lack of updated maps. Mapmakers deal with this difficulty most directly. Navteq, a company that collects road data throughout the country used in popular online map services such as Google and MapQuest, updates its database for Southern Nevada about every three months. Kelly Smith, a spokesperson for the company, explained how even more frequent updates might not solve the problem: "If we were to shoot a map today, it would be wrong tomorrow." The Clark County GIS Management Office updates its database of valley streets on a bimonthly basis, and mapmaker Troy Plocus updates his Las Vegas Fire and Rescue maps just as often. A spokesperson for the Metropolitan Police Department, which uses Plocus's maps, complained that even with such frequent updates, police and firefighters can never have a complete picture of roads in the valley: "Every map service in the world has trouble keeping up with the growth in Clark County."[6]

According to the Regional Transportation Commission (RTC) of Southern Nevada, a countywide public transit and transportation planning agency, more than a hundred new vehicles joined the rush each day during high growth periods. Such an influx led to a 157 percent increase in volume on the county's roads between 1992 and 2008, a rate higher than population growth (133 percent) for the same period. Motorists in the county went from driving 12 million total miles on local roadways in 1990 to an estimated 33 million in 2008.[7]

High vehicle volume, of course, contributes to increased congestion. According to a Texas Transportation Institute report, a typical Las Vegas commuter will spend an average of thirty-two hours every year stuck in gridlock, a more than sixfold increase from 1982. That extended commute, compared to a much shorter nonpeak travel time, makes Las Vegas the seventh most stressed city for commuters, tying it with Chicago and ranking more stressful than nine of the other fifteen largest cities in the country. The busiest intersection through which local drivers pass, the "Spaghetti Bowl" where US 95 and Interstate 15 meet, is considered to be the second worst intersection in the country for freight transportation delays, totaling nearly three hundred thousand hours per year. And, according to a 2007 survey of local business professionals, 90 percent of respondents identified congestion as the greatest barrier to economic prosperity in the region, more than half of them claiming road conditions hindered their own company's progress.[8]

ROAD RAGE

The constant addition of new roads, a rapidly expanding local vehicle count, and an ever-worsening gridlock each has an unquestionable effect on local lives. My interviews provide direct insight into what this impact feels like, from the daily commute on the interstate to arduous trips through a crowded maze of surface streets between home and the neighborhood grocery store. What follows is a sampling.

Several interviewees complained about the congestion. When I asked Jonas and Cora Salvador, who came to Las Vegas in the 1950s and now live in the northwest part of the valley, how growth has affected them, Jonas replied, "Oooh, it's too big." Cora added: "The traffic. . . . Our daughter lives on the south end of town. . . . We don't go see her very often because it takes too much to get there. It's thirty-one miles across town, so we've just decided that we only will see her once a month or so." (As evidence of the always-changing road landscape, Cora asked for help from her husband on what cross streets were near where her daughter lived.) Jonas jumped in to say: "Really it's just the uncontrolled growth and the traffic it causes. And now you get those people that just want their own thing . . . the selfish-type person driving. People are just not friendly anymore. Years ago that is what Las Vegas was known for, but now you get the New York driver against the LA driver and that little city atmosphere is gone."

Nate Jameson put the gridlock in a different perspective: "When I came here in 1965—and you probably remember this too—you could go anywhere in town in fifteen minutes. Now you can't get out of your driveway in less than five. You have to plan for twenty-five minutes or a half an hour to get places, and even then you'll probably be late." Jameson lives in a more central location than the Salvadors, so his travel times are likely shorter than those of suburbanites making cross-town trips.

Jameson, who came from Los Angeles, compared the changes he sees in traffic congestion to his former home. He told me about his decision to move. He saw the direction Southern California was going and didn't like it, so he took a job with the *Las Vegas Sun*. Later in our conversation, when

Looking east at the intersection of Valley View Boulevard and Sahara Avenue. This is "any intersection" Las Vegas and shows the everyday surface-street traffic that is a source of headaches for many locals. Photo by author, June 2007.

I asked him to describe living in Las Vegas, he replied: "Compared to the 1960s . . . [he paused] . . . if I was to move here now, I'd probably move out." When I asked why, he said, "Well, it's getting as bad as LA, traffic wise." Patricia Joseph, another Los Angeles transplant, agreed with Jameson. Looking back at how bad it's become since she arrived in 1989, she lamented that traffic was one of the things that she thought she'd left behind.

The above statements illustrate how noticeable the changes in congestion and traffic can be to longtime locals. "Chopper Tom" Hawley, a reporter for local television station KVBC who has been covering transportation issues for many years, explained this phenomenon: "One thing about Southern California is that they're . . . used to really high volume traffic . . . whereas in Las Vegas, it's a more recent phenomenon, so the people who live here aren't used to it. Under that hypothesis—which is not tested—if you ask longtimers, you'll get a response that traffic is a lot more horrible."[9]

Other more recent transplants, such as Renae Shaw and Sandra Peters, complained about the city's congestion *and* its discourteous drivers. Shaw explained how she lives two to three miles from work, but even if she leaves home at 7:30 A.M. she likely will not make it to the office until 7:55. Then she described her impression of motorists in the city: "I don't know where they get their driver's license. [I think the problem is the] influx of people from out of state who haven't bought into the Vegas community sense. I won't go out on the freeway after, say, 3:00 P.M. I know that if I do, [I might die and] wake up in heaven. I know I'm not going to hell, so I'll wake up in heaven."

Peters explained her "two pet peeves about Las Vegas," both of them related to traffic. First, she blamed city planning for problems on the roads: "Who was stoned and high when they planned this city?" Then she ardently explained her second gripe: "Las Vegas has *the* worst drivers in the nation! I think it is partly because of all the people from different parts of the country with different driving habits. . . . And I-15 is a major thoroughfare between California and Utah so you get all the truck traffic." In sum, she said: "Our streets are terrifying. I've never driven in a place where so many people have cut me off. When we moved here [from San Francisco in 2003] our auto insurance went up 15 percent. The agent told me, 'you will get in an accident in the first year.' I didn't get in an accident until I had been here three years, but then I had three in a row."

Other locals compare their experience here to places they lived previously. New Jersey native Father Kent Sharp, who came to town around 2001, claimed that such behavior was worse than in other cities with stereotypically bad drivers: "I've driven in thirty-eight countries. I drove in New York and New Jersey, and Las Vegas has *the* worst drivers I've ever seen—except for maybe Moscow. I *know* New York driving too!"

Beth Foster, who came to Las Vegas in 1956, doesn't like what the city has become, particularly on the roads: "Everybody drives like idiots. People just pass you like you're crazy even if you're going seventy miles per hour." I chuckled as Beth told me this because of an experience I had on my way to our meeting at her home. I was driving west on Craig Road, a busy street in the northern Las Vegas Valley. As I moved along in the right lane adhering to posted speed limits, a driver in a blue sedan inched up to my bumper, honked, made a hasty lane change, and blew furiously past me before rushing back into the lane directly in front of me. Apparently he was in a big hurry and wanted the right lane all to himself.

I like Al Zanelli's thoughts on Las Vegas drivers. When he used to drive a Citizens Area Transit (CAT) bus (now RTC Transit after the Regional Transportation Commission), he noticed how people sped away from a traffic light only to stop at the next one to wait. He always wondered where people were going in such a hurry. He added: "See that scooter out there. That's how I get around." I responded: "Wow, you take your life in your hands, don't you?" Zanelli continued: "I have to. Gas is too expensive. . . . Yeah, they're crazy out there. In such a big hurry to get home. . . . I never am. . . . They like to mess with me . . . push me off the side of the road. That thing [pointing to the scooter] used to have a windshield." Al gave me a look, as if to say, "If you know what I mean!"

Letters to the editor add color to the conversation. Newcomers are typically the most vocal, and many of them, like my interviewees, compared their experience to roads in their former home. Tom Wagner did so in blunt, chauvinist remarks: "Having driven through most of the large cities on the East Coast, it didn't take me long to realize that Las Vegas has some of the worst drivers in the country. From the 24/7 drunks and drugheads, you are taking your life in your own hands when you venture out on the highways and byways. We have females who can't handle a car at 65 mph while talking on their cell phones passing us in their gas-guzzling SUVs at 85 mph! No fear in those ladies!"[10] Edward Murphy had a similar opinion: "In my adult life, I have lived in San Diego and in the Bay area, where I commuted to downtown San Francisco for twenty-five years. Nowhere in the entire West, in all my travels, have I ever experienced more out-of-control drivers than I found after moving to Las Vegas two years ago."[11]

In a number of instances, letter writers suggested that drivers with such disregard be dealt with appropriately. Newcomer Tim Webster, for example, observed: "This valley is filled with careless, self-absorbed, rude, and overly aggressive drivers. Not a million-dollar news flash, I know. . . . You'll never be able to halt this madness completely, but a severe reduction is desperately needed."[12] Quentin Aukeman's recommendation for such a reduction, like that of many locals, was better law enforcement: "Obviously, red-light runners are . . . dangerous and should be dealt with harshly. . . . Police officers are needed to strictly enforce those laws, and courts must punish the offenders."[13] In a comment that satirizes Webster's proposal, columnist Tom Gorman advised an extreme solution: "The worst thing about living here is the high proportion of moronic drivers. . . . If anti-car missiles were legal, my Camry would be a heavily armed cruiser. I'd put a gun turret in my sun roof. Kapow! Blam! Take that, sucka. And there'd be one less motorist darting chaotically across traffic lanes, running red lights and driving down the center median or shoulder of the freeway as if it were his personal lane."[14]

I characterize Gorman's comments as sarcastic, but local motorists' feelings often reach such extreme levels. Every so often in the local paper, for example, a news story or particularly pointed observation will spark a back-and-forth discussion within the letters column. Dr. Michael Pravica, a physics professor at UNLV, responded to a *Las Vegas Sun* editorial about how lower fuel consumption could result from higher gas prices as follows: "Alter driving habits to minimize braking and rapid acceleration. I estimate that it costs 13 cents worth of gas just to slowly accelerate an 8,000-pound vehicle (like a Ford Excursion) from rest to 50 mph. Driving in Las Vegas is often akin to driving in a stock car race, with people erratically driving, cutting one another off left and right and racing to the finish, only to end up braking at a red light, wasting inordinate amounts of kinetic energy."[15] John Devine responded to Pravica's sensible recommendation in a way that encapsulates the selfish Las Vegas driver's attitude: "He's right, but it's arguable that this style also would cause you to miss the timed green light, resulting in idling through a red light, wasting the savings. In any event, I would simply ask the good professor and his ilk to please stay in the right-hand lanes when using this style of driving. Thank you."[16]

Bob Coffman wrote the editor of the *Las Vegas Sun* to let readers know how pleased he was to be leaving the city. His comments are so representative of the opinions I heard and read about that I will quote the letter in its entirety:

> Now that I've lived in Las Vegas for two years, people back home in Tucson often ask: So, Bob, what's it like living in Las Vegas? My response invariably starts with, "Well, it is home to the stupidest, most dangerous and irresponsible drivers in the world." And I tell them how my drive across town and back each day, to pick up my wife from work, is what I call "the ultimate video game."
>
> Mercifully, my employer will soon be relocating me to another city. I'd like to offer a few parting thoughts:
>
> • "What happens in Las Vegas stays in Las Vegas": I sure hope that applies to the drivers in this town. Please stay here.
> • In honor of almost half the drivers in this town, UNLV's mascot should be changed to the Red-light Runnin' Rebels.
> • Police departments across the country that can't afford to sponsor defensive-driving courses should just send their recruits here for a week and let them drive around without getting plowed into. Good luck!
> • To those locals who say to themselves, "Whew, I made it" after safely making it home at the end of the day, I remind them they're still not out of the woods. While parked in its assigned space in my apartment complex, my truck was

plowed into and knocked sideways four feet (embedding it in the side of an adjacent sports car) by a Las Vegas driver.

- What is this love affair between Las Vegas drivers and (stationary) block walls? If I ever come back here to live, I'm opening a masonry business.

Lastly, I hope that at least a few of Las Vegas' worst drivers know how to read, and that a few of them might be reading this letter. Because I'd like to ask you: Are you an obnoxious, aggressive, self-centered, irresponsible, law-breaking jerk all the time, or only when you get behind the wheel? Either way, you need help.[17]

Coffman's letter was so pointed, accusatory, and resonant that it prompted five responses in subsequent editions of the paper. I will quote a few highlights. Penny Rice wrote: "Thank you, thank you, thank you! He is so right about irresponsible, obnoxious, self-centered and aggressive jerks behind the wheel in Las Vegas. . . . Mr. Coffman is lucky he gets to leave Las Vegas. I'm stuck here for another year, maybe two (sigh)."[18] Carol Nguyen thought it a shame that someone "who truly expressed my feelings" was "departing our fair city" and added: "I am from Canada, the place where you put on your signal to indicate a lane change and it is like God parting the Red Sea for Moses [as] traffic [stops] to let you in. Not like here where drivers speed up, give you a dirty look, and refuse to let you in."[19]

Phil McKay disagreed with Coffman, insisting on a different perspective: "To read letter writer Bob Coffman's opinion of drivers here, you would be led to believe that this is the only city that has bad drivers. The truth of the matter is that at least 90 percent of the drivers in Las Vegas came here from some other area, such as Los Angeles, New York or, yes, Tucson, which is where Mr. Coffman is from. They learned to drive in these areas and they brought their driving habits, good and bad, with them when they moved here. . . . I say so long, Mr. Coffman, and do not let the door hit you on the back of your car on the way out."[20]

The comments of McKay—that Las Vegas does not have a monopoly on bad driving—illustrate the point of view of some residents that local traffic isn't as bad as people make it out to be. This view is important to consider even though it is not stated as vehemently or often as the complaints. I clipped two such comments from locals who wrote to the *Review-Journal*. R. G. Aldrich agreed that congestion is a problem, but argued that many alternatives to a gridlocked freeway such as US 95 exist in the valley: "Due to the extensive street infrastructure of Las Vegas, traffic here is moderate in comparison to many cities of similar size. There is no excuse to let the apparent traffic problems overcome conservative, and thereby effective driving habits."[21] Jeff West, a Chicagoan who visits Las Vegas often and for long periods of time, wrote a challenge to the Road Warrior: "Good weather and dry roads without giant potholes more than make up for fast drivers. . . . Try driving in Chicago this week and write back!"[22]

Five interviewees argued that traffic is *better* in Las Vegas than most locals say or think. When I asked Ben Wychof specifically about his feelings on traffic, he said it doesn't bother him unless he is in a hurry or late for a meeting or golf tee time. What he does complain about is when a driver moves slowly in the middle of a platoon of faster traffic. Such comments made me wonder if he might be one of those fast-driving aggressive people others complain about.

Longtime locals Don Trimble and Cale (who gave only his first name) both had a difficult time coming up with their least favorite aspect of life in the city. Both pointed out (unprompted), however, that traffic is not it. Don answered my question by saying: "It's not the traffic. Not at all, and I complain about it all the time. It's a product of growth. Sometimes I might not agree with some of the decisions [local planners] make, but they have all this stuff thrown at them and, what are you going to do?" Cale also pinned what difficulties he does see on "growing pains" and provided a logical explanation for the oft-faulted drivers in the valley: "Everywhere you go there will be people from somewhere else that will do things to anger you. It's how you respond to it."

Lynn Park, who came to Las Vegas around 2000 after living in Chicago, San Francisco, and Salt Lake City, agreed with Cale. This city has plenty of aggressive drivers, he said, but you can find *them* anywhere. Even though he dislikes such behavior on the road, Park also believes that city planners are doing a good job with the congestion problems: "Once you drive on the 405 between LA and San Diego with four lanes of traffic packed, you realize that we don't have it bad here."

New Englander Adam Morelli also felt that the intensity of Las Vegas traffic is a matter of perspective. I was shocked when he referred to traffic in his response to my question about what he *liked most* about the city. After explaining how he also likes the many hiking areas around the city, Adam explained that Las Vegas congestion is nothing compared to Boston, where traffic can come to a complete stop for miles around: "It's less crowded than Eastern Mass. Traffic isn't as bad here."

TRAFFIC STUDY

With such an overwhelming number of negative perceptions about local roads and drivers—even considering contradicting opinions—it is safe to say that traffic concerns in the Las Vegas Valley are genuine. Such a conclusion, however, prompts the question, "Why?" Any number of reasons may exist, but I will focus on four I found commonly in my research.[23]

The first and most obvious cause of traffic difficulties, and a familiar theme in Las Vegas, is the city's inability to cope with frenetic growth. In fact, growth is unsurprisingly one of the most commonly cited reasons for traffic struggles. It came up seventeen times in my interviews. Fourth-generation native Russell Busch said traffic is his least favorite part of living in the city as he put growth's impact into perspective: "It's so hard to drive anywhere now. I guess it's the same problem in any big city, but for my whole life this has been the fastest growing city, and for my whole life the government has not been able to plan ahead for that." Tom McAllister, a fifty-something native, had a similar experience: "It's lousy, lousy, and it's not getting any better. . . . I can't remember a time since I was a kid when they haven't been working on [Interstate] 15 or [Highway] 95." Whereas construction on high-volume roads like I-15 and Highway 95 are most obvious, the headaches of construction

A road-widening project on Craig Road at Al Carrison Street (near Durango Drive), cutting off access to one local neighborhood from this major corridor. Local drivers must cope with growth in the transportation infrastructure. Photo by author, February 2007.

zones that McAllister spoke of extend to collector streets and neighborhood roads all over the valley.

Returning to Blue Diamond Road again, we can see a good example of growth-related problems in transportation. The *Review-Journal's* Road Warrior, Omar Sofradzija, said it well: "Route 160 has become the most glaring example of the . . . inability to improve roads to keep up with explosive growth in the valley, where a new resident arrives every seven minutes, a new car hits the streets every 10 minutes, and a new home is built every 17 minutes."[24] Suzan Hudson, the woman who placed crosses at fatality locations on the road, said correctly that vast portions of the southwestern Las Vegas Valley through which Route 160 runs used to be empty desert. Today thousands of homes and businesses are being built in the area, many of them in the Mountain's Edge planned community, bringing increased traffic from construction vehicles and residents, and Blue Diamond Road is the only outlet from the new neighborhoods to other major thoroughfares in the city (e.g., the Las Vegas Beltway and I-15). Furthermore, Route 160 is also the main artery into the valley from the growing bedroom community of Pahrump.[25]

Some observers blame local planning and government agencies for problems on Blue Diamond Highway, which are emblematic of those found throughout the valley. Clark County planners, for example, approved the Mountain's Edge development in 2000, and the Nevada Department of Transportation, the agency responsible for the road, was unaware of the imminent growth at an early enough stage to widen the road when it was needed. And NDOT will continue to play catch-up. At the completion of the first major widening project in 2009, the agency's Bob Mckenzie said, underscoring the seemingly unending efforts by planners to accommodate growth, "And we're not done. NDOT is widening Route 160 from Durango Drive to Red Rock Canyon . . . from two to four lanes."[26]

The funding needed for Las Vegas to avoid complete gridlock is another element of the planning problem. A blue-ribbon task force, commissioned by then-governor Kenny Guinn, reported in December 2006 that $3.8 billion is needed to fill a gap in Nevada Department of Transportation funding for the Las Vegas Valley. By two years later, that figure increased to an estimated $5 to $6 billion with inflation and higher construction costs. Guinn's successor, Governor Jim Gibbons, signed a $1 billion road construction bill in June 2007 after significant discussion and debate in Carson City. The funds were meant to cover only a portion of the state's transportation project needs in hopes of finding a more long-term, comprehensive solution at some future legislative session. Of the bill, Gibbons said: "Is it a silver bullet? Of course, it's not. But it will move us forward."[27] Finding that long-term solution in a state that has a libertarian streak and a legendary aversion to taxes, however, will be a difficult prospect.

Even when planners try to alleviate congestion on the valley's roads with resources they actually have, their efforts often fall short of needs. Jacob Snow, general manger of the Regional Transportation Commission, addressed a common query of local residents: "I'm often asked why we can't keep up with growth. What would that mean here in Southern Nevada? Keeping up would require us to add 2,500 feet of asphalt every day to build roads for those new cars. With each new car requiring roughly five new parking spaces in the valley, we'll need 53 square miles of new parking by 2019." Then he championed other county-based projects that would alleviate traffic through mass transit and efficiency improvements in existing roads. Such projects include rapid transit buses; the I-215 Beltway; and new stoplight-free superarterials like a stretch of Desert Inn Road between Paradise Road and Valley View Boulevard that bypasses the Strip, a railroad, and Interstate 15, a combination of parallel barriers in close proximity that has been called the "Great Wall of Las Vegas."[28]

Coming back once again to the representative Blue Diamond Road, we find additional evidence that growth can be a culprit in both congestion *and* bad driving behavior. On that often-choked highway, motorists entering the road from cross streets often have long waits and end up making rash, and sometimes fateful, decisions to, in the words of NHP Trooper Kevin Honea, "jump out into traffic, hoping that traffic will stop for them."[29] Such actions were a major reason for lowered speed limits. Nevada's chief traffic safety engineer agreed with Honea as he discussed a recent study of conditions on that road: "It wasn't just speed limits. [Auditors] saw a lot of driver behaviors that were extraordinary.... People making aggressive moves. People, because of the congestions and delays, taking some risks."[30] Stated differently, congestion yields aggression. Or, as resident Daniel Lippolis said, "People get hotheaded when they are inconvenienced."[31]

MULTICULTURAL MOTORISTS

The reality that almost everyone in Las Vegas is from somewhere else leads to a common perception among residents that the traffic difficulties are caused by a combination of bad driving habits from all over the world. Ten interviewees, some quoted earlier, gave this as one reason for the problems in traffic. *Las Vegas CityLife* reporter Andrew Kiraly called the phenomenon "commuter multiculturalism" and invoked an oft-used metaphor to describe it: "If multiculturalism is like a patchwork quilt, then local roads are like a patchwork quilt made of angry speeding metal boxes."[32] John Phillips, owner of a local driving school, used a different metaphor for the situation: "Because we're a melting pot, people bring their bad driving habits, and bad habits are contagious."[33]

Living in the city, I regularly saw reckless driving. As I walked my son to school, we had to traverse a busy intersection controlled by four-way stop signs. One February morning, as the crossing guard prompted us to enter the crosswalk, a car pulled in front of us making a right-hand turn against the guard's signal to halt and nearly hitting us. After safely crossing, I mentioned to the crossing guard that, "There sure are a lot of crazy people driving in this town." She replied: "Yes, but they aren't the homegrown type. [The crazy drivers] are all these people who come from out of town."

Chuck Styles also blamed the out-of-towners. "After living in Las Vegas for fifty-three years, I can tell you there are a lot of good people and good drivers living here. They just happen to be locals."[34] Such comments are reminiscent of words written in 1929 by Al Cahlan, editor of the *Las Vegas Evening Review*. He noted with alarm the increased traffic and "weird driving" of the time, which he attributed mainly to outsiders who came to town following the announcement of Boulder Dam construction.[35]

Blaming Californians for the valley's bad driving is common among residents. Such a claim may be justified, since the majority of non-natives in the city are indeed from that state and since so many other elements of the Las Vegas cultural landscape have been borrowed from the Golden State. Darren Sedillo hails from San Diego. He told me of a time when he was driving around Las Vegas with friends visiting from his home state. Someone with California plates cut them off and Sedillo yelled and cursed at "the stupid California driver." His friends asked, "Wait, aren't you from California?" To which he replied, "Yeah, I know what you are like and now you're bringing it into my town. I'm a Las Vegan now and I don't want that kind of stuff in my town."

Carol Nguyen, the woman who bragged about the respectful drivers in her native Canada, also wrote of her opinion of California drivers: "After two years of living in Southern California and traveling four hours daily on the 405 freeway, I had enough of those rude drivers. Now they're all living here!"[36] Cale, who you will recall was not really bothered by Las Vegas traffic problems, gave his theory anyway. You can always tell California drivers, he said, when they are riding a bumper and not passing because they don't feel at home driving unless they can see up close the bumper in front of them.

The tendency in this discussion of multicultural motorists is to blame others for our problems. Such a philosophy extends the us/them dichotomy and the notion that "our place" is better than others and that any problems in "our place" are caused by others "infecting" it. Carol Nguyen's praise for Canadian drivers and disgust for those elsewhere speaks to that point. As an additional example, take the opinion of Kathy Kelly, a manager of a local driving school, whose husband used to be a police officer in Los Angeles: "I hear people blaming California drivers all the time. Maybe it's the bad Northern California drivers, I don't know, but Southern Californians drive much better." Nevada Highway Patrol trooper Kevin Honea recognized that placing blame with others might be unfounded: "I've lived in California, and I've never seen what I see here. Everyone, myself included, needs to look in the mirror."[37]

Recall that Sandra Peters was one local who felt that bad driving in the city stemmed from the influx of new residents and that heavy traffic on Interstate 15 was caused, in part, by heavy truck volume. She also gave another potential reason for transportation difficulties, one that, although less commonly expressed than the first two, resonates with many Las Vegans. She said it is "because of all the tourists in town. Just yesterday I was on the 15 and had a semi on one side and I looked over to the van [on the other side] and it was a tourist videotaping the Strip while he was driving!"

Realistically, it is hard to blame Las Vegas's traffic woes on tourists since only a handful of them ever venture off the Strip. Add to that the fact that most locals rarely drive *on* the Strip and are therefore not likely to consider that boulevard as they formulate their opinions about the city's roadways. Still, as Peters implied, tourists may have an impact, especially on the segment of Interstate 15 that runs parallel to the Strip a mile to the west. KVBC TV's Chopper Tom Hawley, in fact, confirmed that implication: "This being a tourist town, especially in the resort corridor, you're dealing with a lot more outsiders, and it's human behavior that they don't pay quite as close attention in other people's communities as in their own."[38]

Visitor characteristics collected by the Las Vegas Convention and Visitor Authority (LVCVA) bear out Hawley's claim. Of the more than thirty-nine million people who visited Las Vegas in 2007, 60 percent either used their own personal vehicle or a rental car to get around on their vacation. Free parking at every Strip resort makes such a choice attractive. Much of the tourist vehicle volume, of course, is found in the near-constant gridlock along the Strip, but some of it bleeds onto I-15. One study found that every new hotel room built in the resort corridor adds 2.3 vehicles per day to the interstate. Considering the volume of new rooms expected, that impact could amount to an additional 100,000 vehicles per day over the next ten years. When you take into account the volume of traffic already on this road, the numbers become staggering: I-15 was engineered for around 130,000 vehicles per day; in 2005 it averaged 218,000.[39]

The ever-critical task of getting tourists to the city may also indirectly impact Las Vegas roads beyond the interstate. In the late 1990s the overburdened stretch of I-15 between Los Angeles and Las Vegas, the major thoroughfare for Southern California visitors, was in need of a major upgrade. With 23 percent of visitors to the city in 2007 coming from Southern California and 93 percent of them coming by car, bus, or RV, the Nevada Department of Transportation gave $10 million of its own transportation funds to help widen the interstate between Barstow and Victorville, both in California. In addition, Nevada congressional officials worked with their California counterparts to contribute another $24 million of the Silver State's federal highway allotment to California portions of the project. In other words, much-needed money for local road improvements went to California to support a better "commute" for tourists to the Strip.[40]

The prevalence of tourism in the city also impacts valley-wide traffic indirectly based on the hours it keeps. Recall from chapter 2 that nearly 30 percent of Las Vegas's workforce is employed in hospitality and leisure, a sector of the economy that operates twenty-four hours a day, seven days a week. Thus a huge portion of the workforce is splitting its time between three shifts: from 7:00 A.M. to 3:00 P.M., from 3:00 P.M. to 11:00 P.M., and from 11:00 P.M. to 7:00 A.M. What does this mean for traffic? As workers on the early shift leave for work and those on the graveyard one head home, they add volume to the roadway before the rest of the workforce begins their "normal" commuting activity. The same thing happens in the evening as the early shift ends and the swing one begins. In essence, a 24/7 schedule increases the length of the morning rush hour and creates a miniature rush hour midafternoon starting around 2:30. The end result: more time during the day that surface roads and freeways are packed with cars.

Eloise Freeman and I discussed the three-shift nature of local traffic. She quoted her former husband on the subject: "[He] used to say . . . 'Nobody in this town works, they just drive around.'" Then Eloise made an interesting observation that may cause some locals to think twice before complaining about gridlock: It could be a lot worse. "If [the city] would convert to a normal forty-hour workweek, with evenings off," she continued, "we'd really be in trouble. The roads couldn't handle it."

That twenty-four-hour schedule of the Strip also has a multiplier effect. For example, it influences operating hours for locally oriented bars and casinos, which in turn puts a continuous stream of alcohol-influenced people on the roadways. Citing that impact, the *Las Vegas Sun*'s David Kihara explained: "Drunken drivers are just a way of life here in the valley."[41]

A final theory commonly posed by Las Vegans about their roadway tribulations is based on culture. The city is full of people who look out only for their own interests, the argument goes, and such people exhibit this same behavior when behind the wheel. Road Warrior Omar Sofradzija cited this factor in responding to a letter from a recent newcomer who expressed genuine fear for her life on local streets: "That's no surprise to jaded old-timers who lament but grudgingly accept that our roads have become a bizarre Darwinian world where anything goes when jockeying for the fast lane. . . . Welcome to the Wild West, newbie." One of the reasons he offered for such a world was "our anything-goes, me-first lifestyles."[42]

Another *Review-Journal* columnist, Jane Ann Morrison, agreed. After describing an automobile crash in which a driver in a sport utility vehicle ran a red light and slammed into a Jetta carrying a pregnant woman, Morrison wrote: "It's the driver's arrogance that makes me wild, the sense of entitlement that red light runners have that their time is more precious than someone else's life. . . . It's the second time I've watched someone blow a red light and hit a car that had the green light."[43]

The comments of other locals further illustrate such frustrations. Devin Strong told me: "Las Vegas has the most atrocious bunch of drivers in the world! People drive as if they are the only ones around." Similarly, Bob McCray's complaint was about driver attitude: "I can deal with traffic, but the way that people drive here. . . . They don't realize that they are dealing with a 4,500-pound machine that can kill someone." Tracy Snow's opinions were in line with those of the *Review-Journal* columnists: "People here are . . . how should I say it? . . . People here are angry. There's not much hospitality. I'm used to the Midwest. I don't like the drivers. They are crazy. That's probably my biggest dislike."

As I drove to and from appointments I became convinced that comments on selfish and greedy behavior on the roads were not limited to people behind the steering wheels. Drive on any major surface street in Las Vegas and you are bound to see more than one jaywalker sprinting across six lanes of traffic in the space of a mile or two, and usually within a short distance of an established crosswalk or traffic signal. The image of a mother pushing a stroller and holding a young child's hand while crossing Desert Inn Road—with three lanes each way and a center turning lane—is a stark and frightening memory. I witnessed such risky behavior more than once. Tim Webster wrote the following to the Road Warrior: "To the jaywalkers, especially at night, think about walking at the (stop) light. I know it's inconvenient, I know. I feel bad for you. I'll feel worse when I read about you in the *R-J*."[44] Rare, indeed, does a month go by when I don't see one or two stories of a jaywalking pedestrian death in the pages of the local newspapers.

In 2006, pedestrians accounted for 32 of 163 traffic-related fatalities within the Metropolitan Police Department's jurisdiction. As a result, Metro ramped up efforts to curb the trend. Police wrote 893 total pedestrian-related citations that year, 570 of those to pedestrians themselves (as opposed to motorists who did not yield to pedestrians in the crosswalk). By the middle of July 2007, Metro had written 2,950 pedestrian-related citations. Still, according to a 2011 study by Transportation for America, Las Vegas ranks as the sixth most dangerous American city for pedestrians.[45]

Officials at local planning agencies and police departments agreed that selfish and disrespectful behaviors contribute to the city's dangerous road conditions. I spoke to Amy Talin, a public information officer for the Regional Transportation Commission, and Murali Pasumarthi, director of the Freeway and Arterial System of Transportation (FAST), about traffic in the valley, including the timing of stoplights to alleviate congestion for which FAST is responsible. Pasumarthi explained that the perception among residents is that they should be able to get where *they* need to go more efficiently. Talin added that the reality of signal timing is that it aims to keep *all* traffic moving in an efficient manner. Part of RTC's plan, they explained, is to get people to use mass transit options so that they can move people, not just cars. But driver attitude makes that difficult too. Pasumarthi explained the self-centered view of many Las Vegans: "It's my car, my road, my freedom to go wherever I want, whenever I want."

Police officers see the behavior firsthand. After making his comment about the number of "moronic drivers" in the valley (quoted earlier), columnist Tom Gorman wrote: "It's not just me who thinks that. The North

Las Vegas Police Chief, Mark Paresi, testified . . . that based on his 32 years of experience, which included driving in every big U.S. and Canadian city, nowhere are drivers more aggressive and disrespectful than they are in Clark County."[46]

One class session of the Metro Citizen's Police Academy I attended was devoted to traffic concerns. Gene Carney, a Metro traffic officer, visited our class and explained the behavior of drivers he has observed in his years on the force. (The fact that he used the term "behavior" several times in his lecture to describe the city's traffic problems is telling.) When a class member asked about officers hiding out of open view in order to "catch" speeders, he shared an experience that illustrates the single-minded attitude of motorists: "Honestly, in this town, you could be sitting out in the middle of the street." He explained how several officers were clearing up an accident and had traffic down to one lane. Drivers were speeding by the accident at seventy miles an hour, twice the posted speed limit, and he decided to begin stopping people. He asked the speeders, "What are you doing?" to which they responded, "I just figured you were busy."

Officer Carney's experience illustrates that many motorists think getting stopped by a traffic cop will never happen to them. In fact, many observers, like Erin Breen, director at the Safe Community Partnership in UNLV's Transportation Research Center, blame a lack of law enforcement for the problem: "People in Las Vegas drive selfishly, like they're the only ones on the road, and the reason is they have the perception they're not going to get caught. . . . People see other drivers do terrible things and watch them not get caught, so they think, 'I'm not gonna get caught either.' Everyone lives in a 'not gonna . . . happen to me' world."[47]

But, as Officer Carney explained, rapid growth has affected their ability to enforce the law ideally. Because they have to attend to so many accidents, for example, officers do not have time to proactively stop people who are committing violations that cause such accidents. "The town is outgrowing the police department and we haven't caught up," Carney said. "I don't know that we will catch up." Metro and its fellow agencies in Henderson and North Las Vegas have only one to two officers for every thousand residents compared to the national urban average of 2.5 officers for every thousand people. Even though a 2005 sales-tax increase will help valley agencies hire more officers, Carney added that it's still going to come back to driver behavior: "It's going to take the public's help to change."[48]

By blaming the valley's road problems on driver selfishness, residents are essentially blaming themselves and doing what Trooper Honea suggested by looking in the mirror. But then again, we are talking about selfish behavior; so what Las Vegan is actually going to say outright that he or she is at fault for the bad driving? I, for one, would not admit to being one of the "atrocious" drivers. Yet, in all honesty, I notice my driving style shift when I return to my hometown. It's almost as if a switch is flipped as I drive across Railroad Pass and into the valley. I become more defensive, more aggressive, and drive faster than I would on the slower-paced roads of the Midwest where I now reside. In a way, that shift is one of the things that makes me feel at home. After all, I learned to drive on these roads. Maybe such behavior is ingrained in me. As I continued my discussion with Officer Carney after that evening's class I told him of my experience. He replied that you almost feel like you have to change your driving style just to keep up with things. He used the example of throwing a stray mouse in with others bustling along as a group. That mouse is not going to sit back and watch them go; he's going to run with them.

EXIT RAMP

So far, I deliberately have not separated two major aspects of traffic: congestion and driver behavior. By making such a separation and understanding how this city's traffic situation stacks up against other urban areas around the country, however, we can learn things about Las Vegas that teach us about place more generally.

First, I will focus on congestion. Obviously, Las Vegas is not alone in its high volumes of vehicles on freeways and slow-moving traffic on surface streets. One visit to Southern California will prove that point, especially if the visit involves driving on a weekday. Recall, too, the Texas Transportation Institute figure that Las Vegans spend thirty-two hours per year on congested roadways. That number, while illustrating an important element of quality of life in Las Vegas, does not compare with the sixty-three hours reported for Los Angeles, the forty-seven hours for Denver, or even the thirty-six hours for Phoenix in that same year. By this measure, Las Vegas ranks twenty-ninth among the 439 American cities in the study. But, many locals argue that things will only get worse, possibly approaching the intensity of Southern California's problems, if state and local officials do not take steps now to accommodate the growing city.[49]

Other measures confirm that the congestion concerns in Las Vegas are possibly not as bad as they seem. According to the US Census Bureau, the average one-way commute in Clark County was 24.3 minutes in 2009, lower than Los Angeles's 29 minutes and the national average of 25.2 minutes. In addition, a Brookings Institution study reported that Las Vegas ranked 93rd out of the 100 largest American cities in terms of total annual vehicle miles traveled per capita in 2005.[50] Both statistics indicate that Las Vegans generally live closer to work than commuters in other urban areas.

Turning to measures that more closely follow driver behavior, however, we get a different picture. According to crash data compiled from Allstate Insurance Co. policy claims in 2007 and 2008, Las Vegas drivers were some of the worst in the nation. Although data for Clark County are not provided in the report, valley drivers in the three incorporated cities were more frequently involved in auto accidents than the national average. In the city of Las Vegas, a driver is 21 percent more likely to be involved in a collision than that national average, with one accident every 8.3 years (rank of 144 of 193 examined); in North Las Vegas the accident likelihood drops to 15.6 above the average, or one wreck every 8.7 years (rank 126); and in Henderson drivers will be in accident every 9.5 years, with a likelihood 5.1 percent higher than the average American (rank 96).[51] Of the thirty most populated US counties in 2007, Clark County had the fifth highest roadway fatality rate at 13.52 deaths for every 100,000 people. For fatal accidents where alcohol was a factor, the county ranked third, with 4.36 deaths per 100,000 people. And, for fatality crashes where speeding played a part, Clark County was eighth, with 3.76 deaths per 100,000 people. Road Warrior Omar Sofradzija reported that fatal wrecks after running stop signs and stoplights or from distracted driving happen in Las Vegas at two times the national average.[52]

So, in its traffic and transportation landscape, Las Vegas is once again a place simultaneously the same as and different from other places. The city is "normal" in the sense that its transportation woes stem from rapid growth. And when lined up next to other booming cities, the numbers suggest that city leaders and residents are doing a fair job of grappling with gridlock. Las Vegas, in its congestion, is not all that unique.

The strong feelings concerning gridlock, of course, cannot be negated. What these opinions tell us relates to another aspect of how Las Vegas is like every other place. When anybody talks about their hometown, they often highlight its strengths and exaggerate its weaknesses. We all like to stand out in a crowd, and by advertising the qualities of our place that we think unique (even if they are not), we can do just that. Las Vegans love to brag that their city is one of the fastest-growing cities in the nation. One way to do so is by talking about the horrible traffic.

Still, based on its residents' driving habits, Las Vegas is a unique place. Even though motorists in Southern Nevada may not be "*the* worst drivers in the nation!" as Sandra Peters put it, the numbers suggest that they rank near the top of the list. Such a suggestion leads us to ask why the city's culture has embraced, as one editor penned, "red-light running, speeding and tailgating as skills worthy of merit badges."[53] A discussion of possible answers points to other elements of the city's character.

Perhaps road selfishness stems from Las Vegans always being in a hurry. This, of course, is common everywhere. Present-day American culture is one of immediacy. We live in what my grandfather, English professor and armchair philosopher, has called "the microwave generation." We want everything now, and that desire is reflected in our driving. Such behavior, in the words of former Las Vegas mayor Oscar Goodman, is exacerbated in a fast-growing city like Las Vegas: "Our society is in such a hurry, we've forgotten civility and good driving manners."[54]

When I asked Metro's Officer Carney why he thought bad driving behavior was part of the Las Vegas culture, he echoed the mayor's words: "Everything is fast in this city. We want things now. It's about instant gratification. People sit in a restaurant and if they're not being helped in a couple of minutes they get upset." Maybe this is why the Las Vegas Beltway—the freeway connecting the south, west, and north edges of the valley that would be more aptly named the Las Vegas Horseshoe for its "C" shape—has also been called the Las Vegas 500.[55] Ted DeAngelo, a retiree from Brooklyn, summed up the "in a hurry" attitude of drivers somewhat sarcastically: "[Drivers] cut you off and if you want to merge in, they just speed up. Then I'll think, 'Yeah, so now you're one more car ahead of me.'"

Perhaps Nevada's culture of individualism and "leave me alone" attitude has crept into local driving attitudes and culture. John L. Smith has equated such an attitude on the roads with his least favorite aspect of the city, what he called "the boomtown mentality": "It's the fast-buck guys that have no attachment. They never vote, they never care, they never let you in [when you're] in traffic." The lack of attachment to place in Las Vegas that I discussed earlier may add to the selfishness on the road. In this way too, Las Vegas is an exaggerated form of a broader American culture.

Perhaps the trait of selfishness and greed on the roads in the local's Las

Vegas is an extension of the stereotypical "Vegas" character epitomized by the gambler. The gambler, of course, thinks only of himself; rarely does he show generosity at the green-felt table unless it is in a handsome tip to the dealer after a big win. If he did show compassion, he would lose and thereby cease to be.

Even dealers are encouraged not to be friendly at the tables. Bernice Grand told me that she initially pursued a career in dealing cards because she wanted to become that cheerful dealer she never had when she gambled. In her first job in a casino, however, she was told to stop smiling at players because the house was concerned that such behavior might indicate cooperative cheating between dealer and player. It seems inevitable that, with so many people involved in various facets of the gambling trade, the inherently greedy behavior of the casino would bleed over to other elements of the community, including driving.

By extension, the unsuccessful gambler shares an additional attribute with the Las Vegas driver: hopelessness. The stereotypical but realistic "down and out with no money for rent" image of an out-of-luck player comes to mind. It is easy to see that such a person would have no regard for other people on his way to the pawnshop to get rent money. *Las Vegas Sun* columnist Tom Gorman asked UNLV psychology professor Christopher Kearney why people drive so rudely. Gorman described Dr. Kearney's response as follows: "He told me these motoring morons are petty, selfish, socially unskilled people who have difficulty at home, work and at play. They compensate for being social incompetents by becoming road bullies, which becomes the primary source of their self-esteem. . . . They measure success by counting how many cars they've passed during the day, without regard to the danger they put the rest of us in."[56]

I'll give the last word to my father, who like his father is also something of a philosopher. Only two weeks after arriving to do fieldwork, I already had several near misses on the road. One night after I returned from the grocery store and more than one close call, I made a general remark about the drivers in this town. He said that, after thinking about the behavior often on his daily commute, he concluded that it came down to a lack of hope, too many people not having much of anything positive in their lives to hold on to. "So, they just rush about trying to get to the next thing." I just hope the next thing isn't the side of my car.

Locals in a Tourist City

had to look twice the first time I saw it. Every so often a youth leader will wear his Boy Scouts of America uniform to church to show support for an upcoming event or commemoration; that much was not unusual. What surprised me on this Sunday was the council patch on the sleeve of one particular gentleman.

Growing up in Southern Nevada, I was a part of the Boulder Dam Area Council of the Boy Scouts, which served sections of Nevada, Arizona, and California, and our council patch appropriately showed an image of Boulder Dam. Boulder Dam, of course, had such a "scouty" feel. My troop held a number of camping, canoeing, and hiking excursions above the dam on Lake Mead or below it on the Colorado River. My grandfather even remembers visiting the site as a Boy Scout while it was under construction in the 1930s. Boulder Dam was a good representation of both Boy Scouts *and*—by nature of its historical significance—the region.

Seeing the new patch, I was surprised that the council had changed its name to the Las Vegas Area Council. The council still serves the same tri-state area, but the change makes sense given the huge population growth in Southern Nevada and the general acceptance in recent years that "Las Vegas" now connotes more than just sin and gambling. A greater shock came a split-second later when I recognized the symbol that replaced Boulder Dam as the representation of the new council. Given a name change, of course, the symbol had to change too. What choice did the council make? What else but the famed WELCOME sign.

On that day I was deep in thought about the *local's* side of Las Vegas in the initial stages of my research. To see that sign representing Boy Scouts, something that figuratively could not be farther from the Strip and "Vegas," seemed to threaten the entire premise of my work. Like many locals, I sometimes attempt to convince my outsider friends that living in Las Vegas is like living anywhere else. But the new patch said otherwise. However, my initial shock at seeing the new symbol for area Boy Scouts may have been

TOP: The Boulder Dam Area Council shoulder patch I wore as a Boy Scout in Las Vegas. Used with permission.

BOTTOM: The Las Vegas Area Council shoulder patch that Boy Scouts in Las Vegas wear today. Used with permission.

somewhat exaggerated. In the course of waging the argument that Las Vegas is a normal place, after all, we locals typically refer—sometimes explicitly, sometimes implicitly—to the uniqueness of the place, its tourist landscape, and what it represents.

Such was the case in my interviews. Of the fifty-three interviewee responses to my question about what locals might tell an outsider about their city, the second most frequent description (mentioned fifteen times) centered on the claim that Las Vegas is a normal city except for some quirks such as slot machines in grocery stores, movie theaters in casinos, and an all-night culture. An additional ten interviewees described the city similarly when responding to different but related questions. These "quirks" are substantial and provide a subtext to thirty-two other interviewee responses about what they tell outsiders: eighteen said they speak of convenience, six touted economic and quality-of-life opportunities, four gave advice on gambling, and another four tell jokes related to life in Sin City. For locals, a connection to the popular image of "Vegas" is difficult to avoid.

As I observed life in Las Vegas, listened to the words of interviewees, and pondered what I heard, it became ever more apparent that the choice of the WELCOME sign for the council patch symbol made sense. It is a recognizable and realistic representation of the community in which the region's Boy Scouts serve and play, even if it isn't as "scouty" as Boulder Dam. Although the Las Vegas of locals *is* different from the tourist realm, it is impossible for anybody to completely disconnect themselves from "Vegas" and the image represented by the WELCOME sign.

So far, I have noted a number of qualities that explain or imply such a connection, but typically with the caveat that these traits may simply be more prevalent in Las Vegas. Here, and in the following two chapters, I will focus on the overt uniqueness of the city from the local's perspective, which is, perhaps, the most intriguing part of the place and its personality. After all, illustrating connections between the tourist place and its local counterpart harks back to my opening argument that Las Vegas is two places sharing one space. Such proximity is bound to foster cultural cross-pollination. Although it may be hyperbole to suggest that some characteristics of the city are particular to this place—especially with the spread of gambling across the country and world since the 1990s—it is also true that the portrait I sketch in the following pages would be difficult to apply to other locations.

LOCALS AND THE STRIP

I want to add two more items to the running list of local Las Vegas clichés quoted in previous chapters. I've already stated the first, the argument that life in Las Vegas is normal. The second has similar undertones and is heard just as often: "I don't go down to the Strip unless family or friends are in town." I heard something akin to this phrase more than a dozen times in interviews.

Almost all locals, it seems, need an excuse to go to the tourist corridor. Flint Salvador remarked: "I don't go to the Strip, unless there is a specific reason, like a show or something. We like to go to the Bellagio for the flower exhibit." Chuck Ballard said similarly: "I don't go down there unless it is to see a show or go to a black-tie dinner." Master plumber Ted Burke explained that he once went to Caesars Palace to attend a convention for his trade. Since his tag noted his hometown, a woman came up to ask where the restrooms were. When he answered that he didn't know and that this was his first time in this casino, she replied: "You live here and you don't come here?"

Other locals avoid the Strip even when they have visitors. Jimmy Del Toro, a cook at one of the megaresorts, explained how the Strip gets boring after you've seen it so many times. When family comes to visit and want him and his wife to act as tour guides, Del Toro says: "Take the keys and the car and we'll stay here." Shari Nakae used to enjoy going to the Desert Inn and some of the smaller, now-extinct casinos, but noted: "The Strip is there if I want it. When do I go? When I have company. And even that only lasted for like five years. After a certain point, I just gave them a key and told them to have fun." On the rare occasion when she does go, her trip is deliberate: "Whatever I am going there for, being a local, I go directly to it. If it's a show, if I want to see the lions at MGM, I just go see the lions. . . . Once or twice a year, if I want to hear good music and enjoy the fountains, I'll go [to the Bellagio] in the evening."

Newcomers and locals who work on Las Vegas Boulevard form one group of exceptions. Tracy Snow and Adam Morelli each explained that they did what many recent transplants do until "playing tourist" gets old. Adam said: "When we first came, we would go once in a while. Now it's old hat and we don't go that often. We go when we have visitors" and occasionally for a Sunday lunch when it's not that busy. Mavie Roberts, a political fund-raiser, said: "When friends come to visit or when I have [a political] event, or a special dinner, that's the only time I go to the Strip really." Mike Bridger spends a lot of time schmoozing clients on the Strip because of the vodka company he owns. He recapped his week for me on the Boulevard: he people-watched (a popular pastime for many locals) as he attended a wine tasting on one day; he saw a Stevie Nicks concert on another; he planned to see a play on the day he and I met; he planned to attend a country music concert and then have dinner at Benihana in the Hilton (rebranded as LVH in 2012) the following night. Bridger said this is a typical week for him because of what he does for a living. For the regular Las Vegan, however, hanging out on the Strip is rare.

It wasn't always that way. In decades past, visits to the Strip were everyday events for local individuals *and* families. Several longtime locals recalled experiences with the tourist corridor from that earlier era. Annie Abreu, whose father came to the city from Cuba in 1953 to work as a dealer, provided one valuable glimpse. Abreu explained that the biggest change she has seen since her childhood years is in the city's casino culture: "When the Stardust went down, people from my generation almost died." She recalled how parents often took their children to the casino. For children of divorced parents, a day with dad often involved going with him to work at the casino, which, she said, would be great fun. "It [the Strip] used to be part of the family."

Eloise Freeman's experience growing up in Las Vegas forty years ago gives another taste of a bygone era. She told how her Brownie troop used to borrow horses from a nearby stable and ride, without chaperones, to the Strip to have breakfast. They left the horses at the Old Frontier, walked through the adjacent Silver Slipper's back door, and then through the casino to the dining room. Nobody thought a thing about it, she said. Her mother and father were not directly involved in the gambling world, but she always had a friend whose father was a pit boss or higher in the casino ranks. She noted how they would often get "comped" for food and entertainment in the casinos. Eloise said, "We went there more than [to] the pizza parlor."

I should explain two important terms in the Las Vegas vernacular related to this subject. The "comp," or complimentary pass/ticket, is a vital piece of Las Vegas culture, today and yesterday. As Brian Frehner explained it, "'comp' functioned as either a noun or a verb. You could receive a comp . . . or someone could comp you."[1] The other term known to longtime Las Vegans is "juice." If you have "juice" you have access to comps or access to someone in the casino with the ability to comp your meal, show ticket, or hotel room. It is both "knowing a guy" and "knowing how to play the game." Eloise had juice based on the people she knew.

Bonnie Pratt told another Vegas juice story about locals and the bygone Strip days. Pratt taught school and tutored in the evenings (she needed the money, as she was raising her three daughters as a single mom). One evening she was visiting with some friends at the school when one gentleman suggested they all go down to see a Frank Sinatra dinner show after work. She recalled: "They thought they could just go down there and get tickets. I was the one without the money then, so I just went along with them. When we got there, the maître d' saw me and [remembered that] I had tutored his children. When he saw me he came up and we got a front row table and the whole thing comped."

In the past, the Strip and downtown were something of a playground for local teenagers too. Trish Allison explained: "Our idea of 'wild and crazy' was to dress up in togas and go down to Caesars Palace." Brian Frehner told of summertime adventures of the same sort. As a twelve year old, for example, Frehner and his friend would ride their bikes down the Strip. "It was tradition for Robbie and me," Frehner wrote, "to stop on the sidewalk in front of Caesars Palace, where at the base of the hotel's marquee were life-size statues that included everyone from Caesar himself to his Roman servants. As tourists walked by gawking at the Roman 'art,' Robbie and I involved ourselves in conversations with Caesar and his retinue. . . . Most passersby quickly understood that two kids were having fun at their expense, but occasionally some walked away with a look of disgust, holding firm to the conviction that they had beheld what happens when children grow up in Sin City." Then Frehner and Robbie would enjoy a gourmet comped breakfast at the Aladdin, where Frehner's mom worked as a secretary in the casino's executive offices, as well as an all-access tour of the casino's "Eye in the Sky" security headquarters.[2]

Cruising was another pastime for local teens. In its earlier years, Fremont Street in downtown Las Vegas was one such route. Misty Carlton explained that, as a young child, social clubs, swimming, and watching the atomic bomb testing (her teacher kept children inside until the flash was gone and then let them go out and observe the mushroom cloud) took up the majority of her free time. "When we were older," she continued, "we drove up and down Fremont Street." The father of one of my high school friends often told stories of drag racing on that road in the 1950s. Fremont Street was once the "Main Street" of Las Vegas and resembled any such street in America (plus the local touch of several casinos), including its cruising culture. No cars are permitted on the core block of the street today, which is now covered by the lighted canopy Fremont Street Experience.

The Strip was the target for the next generation of teenage cruisers. Bryn, who came to the city in 1978, reminisced about a shorter, uncongested Las Vegas Boulevard: "Oh, you know how you could just go down and cruise the Strip. Back then you would get to Flamingo and there was nothing, so you'd just bust a U-turn and go back. Now, you don't even know where Flamingo is." Aric Walker explained how parks and other recreation amenities in the city make a difference for children in the valley today: "There's so much for kids to do now. When I was in high school I didn't have that stuff. We cruised the Strip, did some illegal drag racing, and went down to the race track." Of course, cruising on the Las Vegas Strip today is nearly impossible. It may take more than two hours just to drive a three- to four-mile stretch.

Shannon McMackin also cruised the Strip as a teenager but played there as a child too. Before reaching driving age, she wrote, the Strip was "an amusing baby-sitter." When her parents had tickets for a show, they would leave their two children to play in the MGM's arcade or at the promenade at Circus Circus, one of the only resorts before 1990 that contained anything resembling family fare. But McMackin, who grew up in what was then the outskirts of the urban area, asserted, "when someone asks where I am from, the Strip doesn't immediately come to mind. . . . [It] was a distant landscape, far removed from the empty desert that encircled our house. Yet we went into town and utilized the amenities of the Strip—we swam in nearly every hotel pool, dined at the buffets, and played in the arcades—just as any family uses Main Street in Anytown, USA. That's what these places are for!"[3]

In addition to cruising, some locals indulged in more mischievous types of recreation in the tourist corridor. Walker told me of sneaking into downtown casinos to drink and gamble as a teenager: "When I was sixteen or seventeen and we used to cruise on Fremont Street, we would sit out in the entryway to Binion's. Benny [Binion] didn't care about anything as long as you were paying or playing. So, we'd play the penny slots and drink fifty-cent Heinekens." With a twinge of guilt, Don Trimble told the story about how he and a friend got dressed up, drove down to Caesars Palace on a fight night, and snuck into the boxing match and postmatch party. "It wasn't like we got in trouble, but there were only so many things you could do then," he said. "Now that I look back on it, I'm glad we did it because it was a lot of fun."

Trimble's other Strip excursions highlight the normalcy in locals partaking in its offerings. He recalled: "Growing up . . . for prom we would go to the nicest place at the Dunes, and it would be comped." He often had friends whose parents worked at valet parking or the bell desk and would set up the entertainment for these high school formal dates. Trimble explained that the stereotype that everyone in Las Vegas works at a casino was closer to the true local experience back then: "At the time, perception was probably more like reality."

Although I heard many stories of growing up in Las Vegas when it was a much smaller place, I didn't meet many people who would have been adults in these earlier times. From the few stories I did collect from old-timers, however, it seems that their use of the Strip revolved around access to shows. For example, my neighbor Mitch explained how he used to enjoy the production acts at the Strip casinos in years past, when he could see them for free. He remembered a former neighbor who worked for one of the casinos and would get him comps to such shows.

Of course, some connections to the Boulevard still exist today. I occasionally heard stories of someone working the juice to get into an expensive concert or show. Recently retired high school teacher John Okamoto noted that expensive homecoming and prom dates to attractions in the tourist corridor are still common. "It's the Las Vegas touch," he noted. Such was my experience. As a teenager in the 1990s, I took my senior prom date to dinner at Hugo's Cellar in the Four Queens Hotel, one of the nicer restaurants in town at that time, and to what was then Cirque du Soleil's newest show, Mystère, at Treasure Island. But I had to pay for it all—no juice, no comps. Pool hopping in the resort corridor is still a pastime for some locals, but with restrictive room-key-only access, this activity has become more difficult.[4] I know of teenagers who, in an alternative form of cruising, regularly rollerbladed up and down the Strip, sometimes falling into trouble with Metro cops. In general, however, locals today do not have the bond to the Strip they once did.

Most observers tie the decline in local use of the Strip to changes in casino ownership. After describing her experience "playing" on the Strip as a child, Shannon McMackin explained: "This isn't an option for locals anymore. The new mega-resorts have made the Strip inaccessible. . . . Greasing palms, slyly passing a few bucks, that was Las Vegas! Not any more. The 'cha-ching' of coins has been subverted by a bigger payoff at the corporate level."[5]

The culture of comps and juice also has changed. Nate Jameson was nostalgic for that particular trait of yesterday's Las Vegas. A local since 1965, he has worked as a reporter, off-road racer, and PR man for the Mint casino downtown and for personalities such as Mary Wilson of the Supremes, Robert Goulet, and Lola Falana. Jameson explained: "You could do anything in those days." He told me about one regular gambler at the Mint who typically lost $200,000 a weekend, "which was a lot of money in the '60s." The bosses would give this guy anything he wanted. He wanted a Cadillac convertible and they gave him one. "Each executive has a stamp, you know for comps," Jameson continued. "Well, he asked for one of those stamps. We gave him one, and the next thing we know we saw his stamp all over that casino." You couldn't do that anymore, Jameson concluded. As a gambler, you would have to spend big for something like that. Annie Abreu remarked how her father used to be able to easily comp four tickets to a show, but now under corporate leadership the casino bosses hesitate to do so without the receiver holding a $10,000 line of credit with the casino. "Everything is a money maker instead of a perk," she insisted. Brian Frehner summarized it well: "Corporate gaming squeezed the juice out of Las Vegas."[6]

AMENITIES IN A TWENTY-FOUR-HOUR CITY

Even without the relatively close relationship to the Strip that once existed, today's Las Vegans embrace benefits embodied in the tourist landscape. One such set of amenities stems from the city's entertainment and twenty-four-hour culture, on *and* off Las Vegas Boulevard. In fact, the convenience the city affords its residents—in entertainment, food options, twenty-four-hour access—was the characteristic most commonly touted by my interviewees in describing their city to outsiders. The response interviewees gave to another question regarding what they liked most about living in Las Vegas is also revealing. Among the fifty-seven people queried, the most common answer was entertainment amenities in a city that never sleeps. Fifteen additional interviewees praised those same amenities in conversations focused on other questions. A typical response was that of Howard Schwartz: "I can shop twenty-four hours a day. There's always something to do. I'm never bored." Or Ben Wychof, who found the abundant "activity level" sometimes difficult to balance with real life: "Anytime, day or night, there is something to do. I have to budget my time so that I can just stay at home and read a book or do some work. Last night, I could have stayed out with a friend I had a drink with, but . . . Tonight, I'm not going out." Or Russell Busch, who highlighted entertainment options: "I love the easy access to entertainment. On the rare occasion that we do want to see a show, there are so many options."

Map store owner Bernice Grand praised the geographical options in Las Vegas Boulevard's themed resorts: "When those telemarketers call trying to offer some great deal on a two-day vacation to somewhere, I say, 'I have everything I want right here. If I want to go to Paris, there's Paris. If I want to go to Venice, there's Venice. If I want to go to New York, I go there.' There's no place like Las Vegas. . . . Anything you could want is right here." Mike Bridger had a similar response to my question: "Anything you can think of doing, you can do here. You can go to the Eiffel Tower or on a boat ride in Venice." Of course, we could add to Grand's and Bridger's lists ancient Egypt, Caesar's Rome, medieval England, Italy's Lake Como, and any number of tropical paradises.

Busch also enjoys the excellent hospitality that locals can receive in this postindustrial, service-oriented town: "We have great service here, at all the restaurants. When I go to eat out of town, the service is slower. We have the greatest service in the world!" Mavie Roberts concurred when asked about her favorite part of life in Las Vegas: "I guess it's the thing I miss when I travel other places. The level of service is great here. Everything is so hospitality-driven. It's first class. You just get great service." Another perk for residents is reduced local pricing at many of the tourist entertainment venues, provided the person can show a valid Nevada driver's license.

A handful of interviewees, while recognizing a local life typically separate from the tourist culture, look forward to the occasional opportunity to switch roles, so to speak. UNLV professor Ernest Bannister said he and his wife sometimes enjoy performances at Strip resorts: "You can't beat this city for entertainment. We live the life of the tourist maybe once or twice a month, but usually we wait until we have visitors come into town. Then we step out as tourists. There is so much to see and do in this city!" Aric Walker, who grew up here, explained that one of the things he likes about the place is that you can "live the Las Vegas life for a few days and then get out. Anything you want, you can have it." I like Sandra Peters's description, which is a good summary of the connection locals have with the tourist Las Vegas. As we visited at a picnic table in a suburban city park, she used the word "eclectic" to describe the city and talked about how she can be sitting around in sweats and a T-shirt one moment and that night go out on a fancy date: "I do the mom thing, but I can get dressed up and I have the tattoo and the piercing and I can do that thing, and then I'll go back to being a mom. It's all here. It's a good mix. You can't do that in other places like you can here."

Another of Peters's remarks points to additional indirect amenities locals gain from the tourist corridor, namely jobs and tax revenue. She has noticed that look of shock on the face of new acquaintances when they learn she is from Las Vegas. She said: "The first thing I tell them is, 'I don't live on the Strip.' I think that's what they see. This is Vegas [she points around us]. That's the place you work [she points to the Strip]. That's the place you play, on occasion. And that's where the tourists pay for it all." Her statement illustrates a reality for locals: they have a great entertainment resource, a generator of tens of thousands of jobs for locals, *and* a reduced personal tax burden because of gaming and room taxes paid by the nearly forty million tourists who come to the city each year. In that reality, the tourist corridor influences all Las Vegans, not just those who play tourist occasionally.

Some locals, like Flint Salvador and Shawn Newman, choose to separate themselves completely from the gambling aspect of the tourist Las Vegas, but still recognize that it is the main engine that drives the state's economy. When asked how locals relate to the Vegas image, Flint responded: "Well, in general, I don't think the locals can be disconnected from it. Vegas is tourism and tourism is Vegas. Without it, there is no Vegas. . . . I don't think that anyone can say that people in Las Vegas can be separated from the tourism side . . . everyone benefits from it. We have to face that it pays my paycheck too." (Salvador is a firefighter.) Newman saw similar benefits: "I may not like it [the gambling], but I appreciate it for what it is. . . . Without it, we wouldn't be able to build that beltway. It plays an important role in this state." The role of gaming in the local economy becomes even more evident during economically slower times: locals pay close attention to "room rates and gaming takes, restaurant crowds and taxi traffic" in addition to home values. Reporter Bruce Spotleson summarized: "This is one of the aspects of being a true Las Vegan."[7]

Adding more detail about how he personally relates to the Vegas image, Salvador identified the benefits of the unique time dimension in this city that never sleeps: "Las Vegas is a twenty-four-hour town, so there is always something to do. Places are always open." Similarly, Patricia Joseph noted: "It's exciting. You can be entertained anytime you like, or you can do nothing at all." This "anytime" schedule on the Strip is one of the great attractions for tourists seeking escape from the "ordinary" business and pleasure schedules back home.

A number of interviewees confirmed that locals take advantage of the "anytime" schedule, which extends beyond the Strip into residential land-scapes. When asked to describe living here, Cale responded: "The first thing that comes to mind is convenience. We can shop anytime, day or night. You can find places open when you need them." Peter Nickel said that, after cli-mate, the twenty-four-hour convenience is his favorite aspect of life in Las Vegas. Several other interviewees mentioned the all-night availability of everyday things: "I can grocery shop at midnight." Or, "I can get Mexican food at midnight."

But one might ask, "What is so unique about that?" Many grocery stores around the country are open twenty-four hours, fast-food restaurants too. Indeed, I found on several occasions that the perception of "everything open all the time" is either exaggerated or plainly false. As a teenager, for exam-ple, I was not able to get ice cream with friends after a late movie or dance because such shops would close at 9:00 or 10:00 P.M. Some stores in Las Vegas (including the grocery store around the corner from my house) close at 1:00 A.M., and most national-chain, sit-down restaurants close at 10:00 P.M. during the week and 11:00 P.M. on the weekend. Even many of the shopping and dining areas in the casinos are not open twenty-four hours.

At the same time, Las Vegans are more likely to have access to any num-ber of services and products at a late hour than residents of other cities. A number of independent restaurants and diners do stay open throughout the night. In addition, a dozen licensed child-care facilities in the county offer twenty-four-hour service for the night-shift-working parents, some of which are located on casino property. And, with a large volume of workers on over-night shifts, the stores and restaurants that do remain open typically service a larger clientele during the nighttime hours than their counterparts in other cities. In one informal survey, a number of local professionals recognized the city's twenty-four-hour schedule as a unique characteristic of doing business in Las Vegas.[8]

Columnist Jack Sheehan provided a tongue-in-cheek list of what locals can do, day or night, that, although laced with hyperbole, still rings true: "Should you desire, at any given hour in Las Vegas you can do the following: find a bail bondsman, drive golf balls (at a lighted range), get your carpet cleaned, rent a maid, rent a date, rent a mate, rent a stripper, strip a fender, call a plumber, get your back cracked, buy a backpack, purchase a shirt, pawn your watch, get married, get divorced, rent a movie, repair your car, sell your car, rent a car, tow your car, lift weights, lose weight, get a tan, get your cat spayed, and get cremated. You can even get your cat cremated and yourself spayed."[9]

The true uniqueness of Las Vegas's anything-anytime culture, however, is not found in grocery shopping and franchise restaurants, but in the entertainment and restaurant culture connected to the tourist corridor, neighborhood (locals) casinos, and in the local bar scene. Longtime resident Jeremy Mont explained both the unique and mundane sides of the story: "The thing that Vegas does have is the twenty-four-hourness. I love it. It's the greatest thing in the world. I can go to Vons [a local grocery store] at 3:00 A.M. I can eat world-class food until 11:00 P.M. A lot of those places aren't open later than that, but you can find food any time of the night. I love that liberty." He went on to point out that, in many cities, you can't buy a beer after midnight, and bars in other places have "last call," something that doesn't exist in Las Vegas. He continued: "[Vegas] is like the true American town. It's real freedom."

Many interviewees identified the same uniqueness when they compared their experience in Las Vegas to other places they have lived. Devin Strong, a native of Phoenix, said the two places "are much different. . . . Las Vegas is more night-life oriented. . . . When I was in college in Tempe, the bars would close at 1:00 A.M. They would have last call for alcohol at 12:45. Here every bar is open twenty-four hours. Nobody knows what last call is." He added sarcastically: "Recently, Phoenix has modernized and now the bars are open until 2:00 A.M." Mike Bridger, while visiting a friend in another city, heard "last call" at the bar and "didn't know what they were talking about. The bartender said 'that is the last Crown Royal for you,' and I wondered if they were out. I said I would take another [brand], and the guy said, 'No, we have to close.'"

Native Las Vegan Ali Godino realized her hometown was unique when she went away to school at the University of Arizona. She said: "Once, I made the mistake of walking outside a bar with a beer. I was yelled at so quickly. It was, like, 'You cannot do that here!' And, you know what, I feel like you should be able to do that."[10] For Edwin Aponte, another native local, it wasn't just the bars but the access to alcohol at local stores. In South Carolina, where he attended university, Aponte wasn't able to buy liquor on Sunday for his football parties on account of blue laws: "There goes my football Sunday." He missed home at such times.

Aponte identified another trait that he, myself, and many Las Vegans share. Given the easy access to material things in the city, Aponte and I agreed, Las Vegans may be more likely to do things on impulse and less likely to plan ahead. Why arrange a party or a dinner far in advance when you know that everything you need to put something together will be available anytime, day or night. Whether it's purchasing alcohol for a party or an impulsive midnight visit to a restaurant or movie theater, many locals admit that it is easy to get used to the convenience in their town. Another Las Vegas native, Marissa Mendez, succinctly described this cherished local trait: "You can find almost anything to do at any point, no matter what time of day. There is always something going on."

Las Vegas native Drew Barnes identified the benefit, but also the underside of the anything-anytime life: "I've tried living elsewhere, but there's something to be said for being able to do most anything you want, twenty-four hours a day. . . . There's nothing you can't do in this town, even if it is a vice." For some people living the tourist life can lead to the trap of vice. As I discussed briefly in chapter 3, one vice follows another and can lead to a downward spiral that saps financial well-being and more. Such a sentiment is captured in Andre's (no last name given) description of life in Sin City: "I tell [outsiders] it's what you want it to be. There are a lot of things you can do. There are a lot of things you shouldn't do. You can do anything you want, legal or illegal. You make the choice. . . . It's not just peer pressure here. It's the glamorous life. It's being part of the crowd. It's drugs and strip bars. Hey, that's great, but . . ." When his voice trailed off I said, "It can be a trap." And he responded as if from firsthand experience, "Yeah." Darren Sedillo's warning sums up the risk for locals of having access to amenities and temptations in the city: "You can do anything you want here as long as you have the will power." Gambling and sex, of course, make up the bulk of such temptations for locals, a topic I will come back to in the next two chapters.

EDUCATION IN A CASINO COMPANY TOWN

Educational life in this city is directly affected by local connection to the tourist Las Vegas and the twenty-four-hour culture it fosters. The town's three-shift culture, in particular, adds a headache for teachers and administrators. I asked four teachers how their jobs in Las Vegas might be different from that of their counterparts in another place. Each had a similar response: atypical work schedules affect a teacher's ability to interact with parents. Shari Nakae said, "In the at-risk schools, reaching a parent is hard. A lot of times they will take the phone off the hook because they worked all night and sleep when the kids are at school." Bonnie Pratt explained that calling a student's parent didn't do any good. In many cases, bringing up a concern about a student in the middle of the day would just make the parent upset because she (or he) would have to leave work to remedy it. Pratt thought that such a situation was more common in Las Vegas because many mothers work during the night and don't want to be bothered during their hours of sleep. Retired high school math teacher John Okamoto agreed that parents on the graveyard shift are more difficult to get hold of. Kids, in that situation, are more on their own.

For an intriguing indicator of children left on their own, consider Las Vegas's high numbers of "leftovers," or "students stranded on campus because no one has shown up to retrieve them at the end of the school day." The phenomenon is not unique to Las Vegas, but judging by informal surveys of districts in Los Angeles and Chicago, the problem is more prevalent in Las Vegas. Concerned observers point to the city's twenty-four-hour nature, a high percentage of single-parent homes, the community's transience, a lack of extended family nearby, and just plain neglect. One district attendance officer said: "At 4 P.M. we're still looking for a parent for the kid who showed up at school that morning with a 104-degree fever."[11]

My son's first-grade teacher, Ms. Mendez, told me: "The vocabulary is different here. Because of the gaming and the nightlife and Las Vegas being a twenty-four-hour town, the changes are dynamic. Some of the kids aren't used to seeing their parents that much. Mom didn't wake up on time and that's why one is late for school." I noted in chapter 3 that, for Ms. Mendez, "the transient thing is also an issue." She expanded on her experience and the "revolving door" in her classroom with the following story: "Phone numbers change and they don't tell you what the new one is. You might send a note home and the student loses it or the parent never reads it. I had one student [about whom] I needed to talk to the parent and couldn't get a hold of them, so I went to my principal and he had to RPC [required parent conference, an action usually reserved for students with extremely bad behavior] him just to get the parent in so that we could take care of whatever it was."

Ms. Mendez added another characteristic of Las Vegas teaching. She said: "Because of the gaming, it's hard to tell the kids that they need to get an education and that they need to do well in school. They will tell me, 'I can park cars and make more money than you.' I try to explain, 'that's a job, *this* is a career.'" She said that a number of her students see their mom as a maid or their dad as a valet and "they don't see the whole spectrum. 'Do you want to do that for the next thirty years? Is that what you want to do every day?' If it's just the money, then that is what they will choose." She and other local teachers have concluded more generally that the respect for education that exists elsewhere is not to be found in Las Vegas.[12]

It is a fact that Nevada has the highest high-school dropout rate in the country. According to the Annie E. Casey Foundation's Kids Count Data Center, 11 percent of the state's teens, aged sixteen to nineteen, were not in school and had not graduated in 2009, compared to a national average of 6 percent. Similarly, Las Vegas had the highest dropout rate among fifty peer cities that year at 14 percent. Some of this certainly involves an attraction to the workforce, but some is also a result of pregnancy, failure to pass proficiency exams, and so on. And amid difficult economic circumstances, many dropouts say they need to work just to keep the family financially afloat. Also working against motivation to complete high school and go to college, according to sociologist Robert Schmidt, is a labor market in the city that is in need of employees with "good practical and technical skills." Schmidt pointed to the fact that "Las Vegas has the lowest demand for college professionals of any labor market in America."[13]

It is at the college level that the phenomenon of giving up an education for a job is most clearly manifested in Las Vegas. UNLV, for example, reported that only 39.4 percent of entering freshman in 2003 graduated with four-year degrees within a six-year time frame, just above the Nevada average of 35.8 percent, which was the second worst in the United States. That is far below the national average of 55.5 percent for the 2003 cohort and the 51 percent average for western states.[14]

Much of the struggle facing university administrators stems from students splitting their time between work and school or even quitting school because local casinos and hotels provide opportunities to make a living wage without a degree. College of Southern Nevada (CSN) counselors face a similar challenge of convincing their students to complete a degree. Frank DiPuma, director of institutional research at CSN, noted: "We compete with the job market. . . . If your schedule changes, then you can't balance both [school and work] anymore and you drop out." Indeed, Nevada ranks forty-sixth in the nation in terms of an educated workforce, ahead of only Arkansas, Kentucky, West Virginia, and Mississippi. Just over one-fourth of the state's working population holds an associate's degree or higher (only 19 percent have a four-year degree). In terms of workers aged twenty-five to thirty-four, Nevada has the least educated workforce in the country.[15]

Anecdotal evidence supports the statistics. Renae Shaw laments having raised her kids in Las Vegas because of this idiosyncrasy in the local work-force: "The focus was not on education. It was more on a vocation. Why go to school when you can go down to the Strip and deal cards?" Shaw's twenty-year-old daughter is working at a bank right now and, despite being encouraged to get a degree to ensure her growth and promotion with her current employer, the daughter replies that she is not worried about an education. When she turns twenty-one she plans to get a job in the cashier cages in a casino and make $30 an hour counting money.

Nola Arnett said that, while practicing law and taking depositions, she noticed that about one out of every fifteen or twenty people she dealt with did not even have a high school education, let alone a college one. Rabbi David Banks compared Las Vegas to his former home in the Bay Area: "Las Vegas, in general, does not tend to be a supereducated community. In Las Vegas, you can make a lot of money parking cars or bussing tables," so there is a lower standard for education.

The example of the valet parking job came up repeatedly in my discussions. Art teacher and gallery owner Frank Simon commented: "This is not an educated town. You've got Gorman [a large Catholic high school] graduates who could be going to Harvard who would rather park cars at the Trop." As I talked with Baptist minister Dr. Melvin Roberts about the city's lack of educational focus, he noted that someone can make a hundred thousand dollars a year parking cars: "Where else can you find that in any city?"

The six-figure income potential for a valet is actually something of a myth. This is not to say that the tips, or tokes as they are colloquially known, can't be big at times, depending on the vehicle's owner—stories of hundred dollar tokes are not unheard of. And, exact tip incomes, according to two newspaper stories on the Vegas valet trade, are hard to ascertain because of company policies about divulging income amounts or concerns about IRS audits.

Reporters in both articles found that valets likely make between $45,000 and $55,000 a year, although some have been known to make upwards of $80,000. Although valets certainly can live on such a wage, as Anthony Curtis, gaming and Las Vegas tourism expert, told one reporter: "It's a hard job to get, a fairly juiced-in job. . . . That adds to the mystique. You can't just walk up and say 'I want to do this.' You need to know someone, or be diligent."[16]

Even if the six-figure valet income is a bit of a stretch, the job is representative of what many locals turn to instead of college. I gleaned these examples from news stories: Ernie Acevedo nearly finished his degree decades ago but took a job dealing cards at the Horseshoe in order to provide for his family and never went back to school; Felicia Hersch wants to be a museum curator someday, but before she gets the master's degree she needs, she will likely work full-time as a spa receptionist, a job that, she suspects, might be so good that she won't go back to school; Nicole Maturino splits time between CSN and work at a spa and, as such, is extending her time enrolled in school; Jared Rose, a local high school graduate, dropped out of his first year at UNLV to start a company that promotes events at some of the nightclub and bar joints in town; and Nicole Shields doubles as a student at UNLV and a Madonna impersonator at Imperial Palace where she works as a "dealtainer," a person who deals cards while singing and dancing for casino players between hands.[17]

Still, many locals who work in the hospitality industry while going to school feel that completing their degree is important. Each of the last four persons above plans to finish his or her degree at some point, knowing that it will help in career plans. Nicole Shields knows that Madonna impersonation "can't be a permanent thing," and that she will have to turn to her education for a long-term career.[18] CSN history professor Michael Green told the story of one student he taught that sums up the dilemma for many others: "[I'm] reminded of a former student of mine who was a nude dancer who told me once that she enjoyed the work and made a lot of money at it, but she was in school because she knew that, unlike me, she had a shelf life."[19]

VEGAS QUIRKS IN EVERYDAY LIFE

In addition to the twenty-four-hour amenities of a tourist town and their impact on families and education, the Las Vegas image infiltrates into local life in other diverse ways. These are everyday-life things that have the unmistakable and unique impression of "Vegas" embossed upon them.

Work schedules in a three-shift economy affect both lifestyle and a sense of community. Peter Nickel, a pastor at a local nondenominational church, compared his Las Vegas experience with that in Denver, his hometown: "On the block I grew up on, we knew everyone. . . . I can't say that about my experience here. Community is more difficult here generally speaking." Then he gave one of several reasons for feeling that way: "I think it's the schedules here. People are coming and going twenty-four hours a day. You might never see your neighbor because they are sleeping during the day when you would normally meet them. And people work on weekends. That doesn't happen in other places like it does here." Jake Glennon agreed as he explained the lack of community: "I think it's the different working hours. When's the shift change? . . . It's a twenty-four-hour town." People aren't on the same shift, he explained, so they don't have the interaction time in their neighborhoods that happens in other places.

Reporter Marshall Allen gave a name and a face to this anomie. He opened his article with an emblematic neighborhood description: "You've seen it. We've all seen it, and probably given it. The Vegas Wave. You know, that neighborly wave to the couple across the street who you see all the time but never talk to. 'Hey.' *Wave.* Now duck into the car and drive away." Allen continued by positing the predictable question: "We offer that meager, sheepish salute to nameless neighbors to show we are friendly. But are we?"[20]

JR Henson elaborated on the impact of work schedules on community: "We're not the nine-to-five community. A third of the population is working every hour of the day . . . and on weekends. And if you want the tips, that is when you have to be around—on the weekends." Henson's last comment brings up an intriguing oddity in the workforce that is quintessentially Las Vegas. As my friend Jimmy Del Toro reminded me before going to his job as a chef at a Strip hotel, because resort employees often don't work a typical five-day workweek, they are often asked the questions: "When is your Friday?" or "When is your Monday?" Their reply might be something like, "My Friday is on Wednesday and Saturday is my Monday." CSN professor Michael Green grew up in Las Vegas, the son of a casino employee. His description provides another perspective on work schedules in this city that keeps its own time: "My dad was a dealer for thirty years, and my aunt in Phoenix used to say, 'Why don't you come down for the weekend?' My mother would say, 'Your weekend or our weekend?' My dad was off Wednesday and Thursday."[21]

Holiday schedules for workers in the city's hospitality industry are also affected. After an interview on July 3, 2007, with the owner of a helicopter tour company, I witnessed this firsthand. On my way out of the building, one of the company's employees entered the hallway just in front of me and we had a short "excuse me" exchange. He said, "Sorry! Just moseying," to which I replied, "We all should be. Tomorrow is the Fourth." He surprised me by saying: "Well, I'll be right here." Holidays can be some of the busiest days of the year for a business that caters to vacationers taking advantage of *their* day off.

A number of interviewees commented that friends and relatives often request permission to visit and stay with them on their next vacation to Las Vegas. Such an arrangement is, of course, common to all tourist places, and it can be draining. Ben Wychof explained that for nearly every weekend over the next few months he would be hosting visitors. When I asked if such an occurrence was more common since he moved to Las Vegas from Wisconsin, he chuckled and said yes. Trish Allison, in contrast, likes the convenience of not having to travel to keep in touch with acquaintances: "I don't have to go see anybody. They come to see me." And Jimmy Del Toro put this phenomenon in pointed terms: "When you first move here, a funny thing happens. Your house turns into a hotel."

Even without regular visits from tourist relatives and friends, the scene on a typical Las Vegas residential street can take on a distinctive personality. Take, for example, Rio Garcia's description of his Henderson street in the southern part of the valley. The neighbor to the left is a greenskeeper at a golf course. To his right is a computer programmer whose sister, also living there, is a stripper. Across the street is a doorman for the Hard Rock Hotel in one home and an electrician for the city or the county in another. The fact that Garcia didn't pause as he explained all this, particularly when he mentioned the stripper next door, is indicative of the diversity of people fostered in the Las Vegas environment.

If you were a resident in a regular Las Vegas neighborhood, you would see the image of the Strip, gambling, and entertainment manifest in any number of ways. Maybe you live in a neighborhood near the intersection of Russell Road and Maryland Parkway, next to a twenty-two-acre park named after Siegfried Fischbacher and Roy Horn, the famous and long-standing Strip illusionist duo. Rory Reid, a Clark County commissioner, justified the name by saying that: "I just believe they've done a lot for our community. We're a tourist-based community, and when a lot of people think of Las Vegas, they think of them."[22]

Perhaps you and your neighbor plan to participate in the next Las Vegas Marathon. If you do, chances are that you will see a number of Elvises (sometimes given the plural "Elvi") running at your side, or perhaps a couple tying the knot along the way, the Blue Man Group performing on the sidelines, or some other Vegas-only features that are part of each year's race.[23]

Maybe you live in a neighborhood with a school like the one Strip entertainer and proud local Rita Rudner has described: "I know a lot of wild stuff goes on here, but no one ever invites me. . . . We were looking at kindergartens recently for our daughter, and I believe it was the first time I'd ever seen a playground with a stripper pole."[24] Then again, maybe not. Of course, Rudner was being sarcastic when she told that joke after living in the city for several years. But it is true that your neighborhood moms can, for a modest fee, get lessons at one Strip resort on how to dance like a stripper, pole and all, a program that has been quite successful with both local and tourist clientele.[25]

No matter what neighborhood you live in, former Las Vegas mayor Oscar Goodman has become a unique representative for you. In his three terms as mayor that began in 1999, he became such an icon for the city—he was mentioned in twenty-seven stories in six major national papers by a recent count—Goodman has become the de facto mayor of not just the city of Las Vegas but the entire valley.[26] When I asked if he considered himself as such, he told me: "Yes. People call me all the time and tell me they wish they could vote for me." He noted, and I found the same in my interviews, that some people even say they voted for him when, in fact, they legally could not have. He mentioned a specific example from recent memory: "To the people, I'm the mayor de jure. I was the only elected official invited to the groundbreaking of Echelon Place [on the Strip]," which is out of his jurisdiction.

Where else can you find a mayor who built his career (prior to politics) as a defense lawyer for some of the shrewdest mobsters Las Vegas has ever seen; who was the "official spokesman for Bombay Sapphire gin" (he gave his paycheck to charity); who held popular mixers called "Martinis with the Mayor;" who taught a not-for-credit course in 2007 at the Community College of Southern Nevada called "How to Make a Martini with the Mayor;" who told a much younger school crowd (fourth graders, actually, which caused a small controversy) that, if stranded on a deserted island, he would wish only to have his bottle of gin; whose classic pose with gin in one hand and showgirls on both arms has become an image used in national media to represent the city; who has his own bobble-head caricature Vegas-kitsch; and whose business card is in the form of a $100 poker chip?[27]

Even though some Las Vegans criticized Goodman for his often blunt and

The front and back of Las Vegas mayor Oscar Goodman's "business card" in the form of a $100 poker chip. Used with permission.

sometimes outlandish comments, he was an *extremely* popular mayor with locals, as evidenced in his big election wins and in his ability to represent the city at home and all over the world. In a style all his own, he boasted: "I became the brand. . . . When people saw me with the showgirls, they thought 'Las Vegas.' And we used it as a marketing tool, and of course it was fun." Some were saddened when he was term-limited out of office in 2011, because he was seen as such a perfect representation of the city, and for that reason Mr. Goodman will likely remain an important figure in local lore for many years, regardless of who holds the post of mayor in the future. Incidentally, the person elected in 2011 to succeed Goodman was his wife, Carolyn.

Other indicators of the local connection to the Vegas image go largely unrecognized. The Boy Scout council patch that I described in this chapter's opening is one. As I visited the local library with my children, I noticed another: The mascot chosen by the Las Vegas–Clark County Library District for its children's programs is a lizard named Neon, reflecting both the desert environment and the glitzy Vegas lights.

Another such experience occurred shortly after we moved to the city. As my son came home from first grade one day, my wife Rachel, as she normally does, went through his backpack to see what homework, notes, or news he'd brought home. Only that day wasn't normal. From the other room, I heard a shocked, "What the . . ." and went to see what was the matter. Rachel pulled from the backpack a deck of casino playing cards. The cards showed the logo of Sam's Town, a nearby locals-oriented casino. The card corners were cut and the box stamped "L1R1-4-19-06 SWING," indicators that this deck had seen action at the tables during the swing shift on April 19, 2006. When our son inquired the next day where the deck had come from, he was told that someone had dropped off a bunch of used card decks at the school; his teacher didn't know what do to with them and so decided to send one home with each of the kids in the class. "Only here," I wrote in my field journal.

Las Vegas does not have a major professional sports team that might serve as a binding force for a citywide community, but even if it did the team likely would not overcome the dominating image of the Las Vegas Strip. For evidence, consider the Las Vegas Wranglers, an ECHL minor league hockey club, whose fans rarely extend beyond a local audience. Just before the puck dropped to begin the game I attended in 2005, the starting lineup skated onto the ice through a giant slot machine. When the home team scored a goal, announcers played Elvis Presley's "Viva Las Vegas." One style of team jersey shows the player's number in the center of an image of a poker chip; another shows the team logo, which includes a puck in the shape of a roulette wheel. And one Wranglers goalie had a caricatured image of the Las Vegas skyline on his mask.

The Las Vegas image seems inescapable. It is found in locals-oriented advertising and promotion on the radio. In 2005 one station, FM KQOL, carried the tag: "More music from the most exciting city on the planet. Kool 93.1, Las Vegas." Adult classic and contemporary station FM 104.7 KJUL similarly advertised itself in 2007 as: "Playing the music that made Las Vegas famous." And the local National Public Radio station, KNPR 88.9 FM, made a fundraising pitch in 2007 that argued: "The visitors see all the lights, but to you it's home. In a maturing city, culture is vital to the quality of life, which is the mission of KNPR." Such a statement, while recognizing the city's two faces, emphasizes a distancing from the tourist side of the city. But during the same fund-raiser, the station gave away incentive packages to donors that included tickets to major shows at Strip hotels.

The Las Vegas image is evoked in many aspects of city life, but it is nowhere more apparent than in the cultural landscape. The decals on the side of a transport trailer at the local Harley-Davidson dealer, for example, proclaim a slogan that oozes with Vegas style: "Ride all Day, Play all Night."

Another example is the variety of neon and flashy signs in the valley. Consider the sign for one local branch of the Southern California classic chain In-N-Out Burger. You are not likely to see such a display in San Diego or Los Angeles. True, this is a branch location close to the Strip, but it hosts a mostly local clientele. Farther away from the resorts, you'll see a more normal ver-

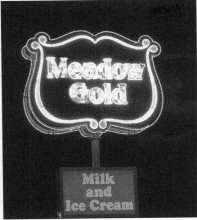

Flashiness abounds in Las Vegas, even in the local's realm. The signs for an off-Strip In-N-Out Burger on Tropicana Avenue and Industrial Road and for a local dairy located miles from the tourist corridor attest to that fact. Photos by author and Dennis Rowley, December 2005 and April 2012.

sion of the hamburger chain's logo. Still, flashiness abounds in the city. Even though its warehouse is many miles from the tourist corridor in both purpose and location, Meadow Gold, an intermountain west dairy company, erected a sign visible from Interstate 15 in the same red-neon and blinking-letter motif used for so long on the Strip.

For another angle on the local landscape, observe the view from the parking lot of the West Las Vegas Library on Lake Mead Boulevard. The iconic Stratosphere is ever present. In fact, when driving around the city, one rarely escapes sight of the mammoth tourist hotels.

My interest in the omnipresent Vegas image was piqued when I heard people, such as Karl Marlin, say that some days they almost forget where they live until they see the Strip and say, "Oh yeah, we do live in Las Vegas!" Sherese Welsh, a sophomore at the suburban Coronado High School, said her favorite part of living in Las Vegas is "the view of the Strip at school each day."[28] Another local called the Strip the "center of mass" for the city. Suspecting that this visual landscape heightens the Strip's impact on locals, I tested my hunch using a geographic information systems (GIS)-based viewshed analysis, which identifies, based on a digital elevation dataset, all the areas

A view of the unique landscape in the Las Vegas Valley. This photograph showing the ever-present Strip landscape was taken looking south from Lake Mead Boulevard and J Street. Photo by author, July 2007.

of a landscape where one can see (or be seen from) a set of defined "observed areas." The map on the opposite page shows the viewshed for the Strip, which contains almost 93 percent of Las Vegas population.

Such an analysis illustrates the pervasive nature of the city's tourist core. Even if Las Vegans say they don't go there, the Strip is part of their cultural landscape and, as such, impacts how they experience the city as a whole. I recall one occasion when I went to a birthday party for one of my child's friends at Exploration Park in the southwestern Las Vegas Valley where the adjacent Exploration Peak blocked my view of the rest of the city. Without the familiar skyline, I had a strange feeling that I was no longer in Las Vegas.

VEGAS PRIDE

More difficult to describe, but no less important to the city's sense of place is the "Vegas" personality that some residents take on. This "Vegas pride" manifests in a number of ways. One of those, surprising to outsiders, is a culture of openness. The Strip, of course, is not afraid of what it is: casinos, topless revues, strip clubs, escort services, and more. That culture bleeds over into the local lifestyle. The acknowledgment that people will choose to participate in such activities reflects this local characteristic. "Do it if you like, at whatever time you like," characteristically nonjudgmental locals often remark, "it's your choice." Oscar Goodman's persona is an example of the city's "truth in advertising" approach. Larry Brown of the Las Vegas City

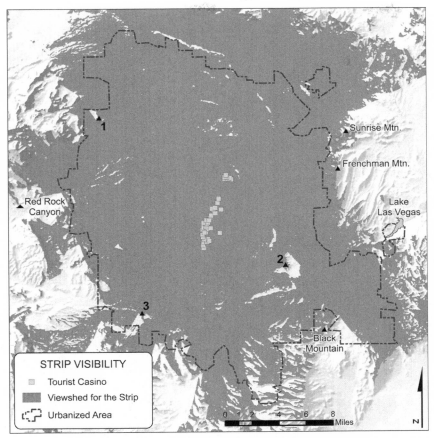

The long shadow of the Las Vegas Strip. The dark gray areas depict locations in the metropolitan area that have a clear view of the tourist landscape. Locations of relief that disrupt a view of the Strip within the urban area are labeled by number. They include Lone Mountain (1) in the northwest, Whitney Mesa (2) in the southeast, and Exploration Peak (3) in the southwest. Map by author.

Council described his colleague: "When he's out, he says and does what he feels and believes. . . . It's refreshing to hear a politician tell it like it is, even if it sounds outrageous at times."[29]

Linda Piediscalzi, coowner of Dead Poet Books, a quirky used bookstore on West Sahara Avenue, described the trait more generally: "We are a community . . . we're kind of different. People work different shifts and have different days off here. It's not like other places. But it's also more honest. That's what I've always felt about Las Vegas, is its honesty. It doesn't pretend to be something it's not."[30] Vegas Girl fashion columnist Emily Kumler echoed Pie-discalzi's perception. One of the great things about the frank, be-yourself environment within the city, she wrote, is that you have the "freedom to dress as wild or tame as you choose at any given moment."[31]

The wild fashion that Kumler noted points to another element of the Strip that carries over into local life: flash and newness. In the years she has lived in the city, Shawn Newman has noticed that, compared to other communities, Vegas has a higher percentage of people who drive late-model luxury cars, wear expensive clothing, or carry high-end handbags: "Everyone drives a brand new BMW or Mercedes. If you're driving a car that's five years old, it's too old. . . . It seems that this place is Diva central. . . . Here it's like the locals have bought into the flash."

Russell Busch agreed: "People dress really nicely here. Everything is flashy: the boobs, the jewelry, the cars." He gave an example to illustrate. When he was in the military, his salary allowed him, as a single person, to buy a red Corvette. "In North Carolina [where I was stationed] people would point at me as I'd drive down the road. When I got out of the Marine Corps, I drove it back here. Let's say the trip is 3,000 miles. For 2,990 miles, there wasn't a nicer car on the road, but for those last ten miles, from Hoover Dam to Las Vegas, it was a dime a dozen." He said it was the same for the number of Mercedes on the road in Las Vegas. Busch added: "That *is* the culture. Your appearance, your car, is more important here."

Locals enjoy being from Las Vegas. That fact alone, it seems, makes you instantly cool and often provokes great conversation. Aric Walker told me: "Here's the thing. I like the Vegas image. That's what I like about being from Las Vegas. I like telling people that I'm from here and seeing their response. I especially like that I'm a native. I can say I was born and raised in Las Vegas and they go, 'What? Really?'" Don Trimble explained: "I've been traveling my whole life and when I say I'm from Las Vegas the guy next to me says, 'Wow, you're from Vegas. I want to go there.' You don't get that when you're from Kansas. You don't get that when you're from New York. Nobody cares." Nola Arnett also travels a lot for work and enjoys the responses she gets when she explains where she is from: "It's so weird [to be] from Las Vegas, because people notice you." For Rafael Villanueva, a sales executive for the Las Vegas Convention and Visitors Authority, being from a city so noticeable makes his job easy: "When I travel, I do presentations to hundreds of people and I love doing it. It's such a great feeling, seeing the attention that Las Vegas gets.

. . . I love talking about Las Vegas. I can't wait to do another presentation in another city. You don't have to sell them. You just have to tell them."[32]

As I listened to Jeremy Mont explain his deep love for his hometown, I realized that he could be the local's equivalent of Mr. Las Vegas, Wayne Newton. Mont knows that when a Las Vegan visits another place he sticks out in the crowd. In fact, Mont takes it to the next level. When he goes out he takes on the Vegas persona—a slick, pin-striped, almost gangster-like suit over a shirt with a long collar with the top two buttons open: "I rock it Vegas. I love it." Sometimes he'll even tease people he meets. When they ask him some silly question about his town, he might respond: "Everyone's a dealer or a showgirl and everyone lives in a hotel." Then he'll add: "I have a house, jackass. People don't live in the coal mines in Pennsylvania." The comments of Mont and other locals about the stereotyped perceptions of their city, although sarcastic, nonetheless show a certain pride that comes from belonging to this place.

PLACE STORIES

Even though many of the stories in this chapter relate to the city's distinctive qualities, the Las Vegas experience can be instructive when considering other places. The Southern Nevada metropolis does not have a monopoly on uniqueness. The first lesson is that every place *is* unique. Locals may argue that life in Las Vegas is as normal as life anywhere, but in reality it is not. Even though locals try to separate themselves from the Strip and what it represents, an inseparable part of everyday life in the valley is tied to the image of a five-mile stretch of Las Vegas Boulevard. In other words, although locals may physically separate themselves from the Strip and what it represents to the outsider, as insiders they still have a difficult time disconnecting themselves from that image.

Sometimes we may think our own place—Lawrence, Kansas; Houston, Texas; Poughkeepsie, New York—is mundane, but each place has its own character. And sometimes it takes leaving the city to realize it. Local librarian Ina Moore explained: "After living here for so long [thirty years], I think that Las Vegas is normal. It's not until I go somewhere else that I realize that it is not." Residents of almost any place often don't realize how much they value that particular locale and how special it is to them until they leave and begin to feel a certain longing or homesickness.

A second lesson from Las Vegas's uniqueness is that the cultural landscape can be a powerful force for place identity and place creation. Every place has landscape features that define it, at least on the surface. Icons such as Central Park or the Empire State Building in New York, the Gateway Arch in St. Louis, and the Space Needle in Seattle perform such a function in these major cities. But even small towns have a distinctive downtown or unique courthouse that can represent that particular place and its character. People's perceptions are often based on the cultural landscape. For Las Vegas the dominant feature just happens to be the Strip, its bright lights, and, by extension, the culture of gambling, entertainment, and the twenty-four-hour life it symbolizes. As illustrated by the fact that so many Las Vegans argue for the normality in their place, we may distance ourselves from a prominent place perception or iconic representation, but usually cannot totally disconnect ourselves from it.

Finally, stories of place uniqueness illustrate hometown pride. I noted many times in my field journal how this or that interviewee exuded a love for their city and how some even said so outright. Annie Abreu said: "I'm a big cheerleader for the city. I love Las Vegas." Trish Allison told me how much she loves her hometown and gave me a charge: "Make it look good." I found also that people were not just willing but genuinely excited to talk about their place. Most people, once they get attached to a location, will likely feel that same pride and that same enthusiasm about their own place.

In the end, the argument that so many locals make to defend their city as a "normal" place is both true and false. On one level, Las Vegas residents live a fairly normal life, involved in the same activities as their counterparts in other cities. On another level, those residents have access to things that their counterparts see only on vacation. On yet another level, that uniqueness is a version (admittedly an exaggerated version) of the uniqueness found in any place.

Life in a Town of Glitter and Gold

often travel by air from my current midwestern home to Las Vegas for a weekend visit. Unlike many of my fellow passengers, my intentions are family- or, more recently, research-oriented. But sitting on a flight inbound to McCarran Airport on a Friday, I always sense the energy of the mostly tourist group. It all starts near the gates prior to boarding the flight. A group of four middle-aged men, for example, joke and discuss the fun they anticipate while celebrating a fortieth birthday in Sin City without the distractions of home, family, and work. Onboard the aircraft, the buzz of conversation is louder than on weekend flights to other cities. Photos are taken in anticipation of the Facebook page update describing the trip. Passengers discuss what hotel they will stay in, what shows they plan to see, how the Strip has changed since their last visit. Alcoholic beverages flow at high rates, too, adding to the energy and decibel level. Descending into the Las Vegas Valley, I have often overheard someone giving the history of this or that part of the tourist landscape to an excited first-time visitor to the city.

I have also noticed a special mood on the return flight Sunday evening. The plane is ghostly quiet. No plans, no laughing, faces are plain and emotionless. Drinks are taken quietly and most passengers sleep. Perhaps they are just worn out from all-night parties in casinos and clubs. Perhaps they are recovering from an abundance of alcohol consumed while throwing dice at the craps table. Then again, maybe they are depressed because of the money they won and then lost on a single hand of twenty-one. This is what Vegas is built on: money left behind by tourists. This is what happens in Vegas and stays in Vegas.

But what about the *people* who stay in Las Vegas? What about the locals? Gambling and adult entertainment make up a large portion of the economy in this city and not just on the Strip. And for many Las Vegans, direct involvement in either industry is a part of life, whether it be as a dealer at Wynn Las Vegas, a stripper at Crazy Horse Too, or a player at the video poker machines at PT's Pub. Granted, as gambling expert Howard Schwartz put it, "the out-of-towner is here with discretionary money. He spends what he has and enjoys the city on a sporadic basis." The local gambler, on the other hand, may end up spending his rent money on a game of poker only to be dealt a bad hand.

The gambling scene is probably the most significant way in which locals are connected to the outsider's image of the city. A close second is the abundance of sexually oriented businesses in this city that is often referred to as an adult Disneyland. Living in a culture of gambling and sex, in short, epitomizes how Las Vegas is a different kind of normal.

AN OMNIPRESENT LANDSCAPE

Gambling by itself does not make Las Vegas as unique as it once did. Today some form of legalized betting exists in every state but Utah and Hawaii. It is also a global phenomenon, the most prominent recent examples being Macau, China, and Singapore, where many of the Las Vegas casino moguls have duplicated their Las Vegas monoliths to cater to a burgeoning and prosperous Asian clientele. Yet, Las Vegas remains the gambling standard. The fact that it is Las Vegas gamers who stand out in the Asian market is evidence of that. So is the claim of local boosters that gamblers living near riverboat or Indian casinos will still be drawn to Las Vegas in order to experience the "real thing."

More important for locals, however, is the omnipresence of the gambling landscape throughout the Las Vegas Valley. Darren Sedillo's response to my question about how the city is unique is telling in this regard: "Obviously, there is the gambling. There's the Strip that everyone knows of, but also on every corner is a 'saloon' where you can spend your entire paycheck in five minutes and maybe get a free beer. Then there's the grocery stores where you see the granny gambler, and the gas stations where you can go in and play the machines."

Of course, hotel-casinos on the Strip and in downtown are the most obvious places to play. But more accessible counterparts located throughout the suburbs—some as large as tourist casinos—cater specifically to the local gambler. Colloquially referred to as locals or neighborhood casinos, these properties offer dining and entertainment options (along with gaming) so ubiquitously that even the nongambler must walk through all or part of a casino to visit a restaurant or see a movie. And they do. Neighborhood casinos are hubs for local entertainment and social activities; nearly every movie screen in the valley and all of the bowling lanes in the city are located within these properties. Smaller venues offer scaled-down gaming opportunities in nearly every bar and retail store. Furthermore, travelers to the city know that every terminal in McCarran International Airport contains rows of slot machines. In short, gambling is embedded in the Las Vegas cultural landscape, an indelible part of everyday life for those living in the city.

The grocery store is perhaps the most unexpected and surprising gambling venue for outsiders who learn about life in Las Vegas. Tracy Snow explained what she tells inquirers: "It's like any other city if you're off the Strip. Except for the slot machines in the grocery stores." I asked how they

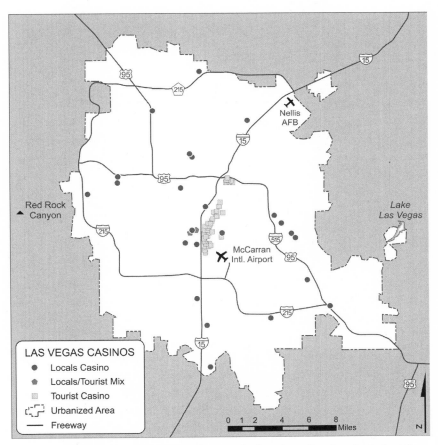

Major casinos in the Las Vegas Valley showing the location of locals-oriented establishments relative to the tourist corridor. Note the classic geographic phenomenon of core versus periphery representing the tourist and local realms, respectively. Map by author.

Sam's Town in 2007. Notice the Century theaters attached to the main casino complex. Photo by author.

respond to that. She told me of one friend who visits occasionally and whom Snow usually meets on the Strip. One time, that friend came a day early and stayed at Snow's home. They made a visit to a local store and the friend was shocked at what he saw: "They really do have slot machines in the stores." A local joke and stereotype of the grocery store minicasino typically centers on bickering children strapped into carts, melted ice cream, or mindlessly squandered grocery money left behind when mom or dad stopped to play video poker on the way in or out of the store.

The habits of locals further illustrate the ever-present nature of gambling in Las Vegas life. According to the Las Vegas Convention and Visitors Authority's (LVCVA) *2010 Clark County Resident Study,* gambling was the second most popular leisure activity outside of the home for locals. Sixty-two percent of residents say they gamble occasionally, and of that number 77 percent do so at least once a month (28 percent gamble at least two times a week). Larger casino establishments are the favorite destination for local gamers: 72 percent patronize such venues, 24 percent play at local bars and restaurants, and 16 percent do so at gas stations, convenience stores, or grocery stores. Based on volunteered dollar amounts from gamblers in the previous edition of the survey (2008), researchers estimated that the adult population in Clark County budgets $3.55 billion for gambling on an annual basis.[1]

Another, less objective indicator of gambling's influence on Las Vegans is found in the prevalence of free-flowing cash in the city. Joseph Mani remarked: "Vegas is about image, glamour. It is largely about having a fat wallet and doing something with that." The "toke"—localspeak for a tip—to a dealer, cocktail waitress, or valet is an outgrowth of the gambling economy that further enhances Las Vegas's character as a cash town. Baccarat dealer Carlin, who typically earns a six-figure income, described his native home in the Twin Cities as an education-oriented town. When I asked, "If Minneapolis/ St. Paul is education oriented, then how would you describe Las Vegas?" he responded: "Money oriented. Everyone is interested in gratuity. That is what my job is. In Minneapolis/St. Paul people are more frugal. Here we are more gratuitous. My dad was surprised at how much tip I gave someone [on one occasion]. We give higher tips here." I myself have taken on that part of the local culture; friends and family have commented to me on my generally higher than normal tips to servers at restaurants.

Most off-Strip casinos offer paycheck-cashing promotions, a phenomenon that infuses huge amounts of cash into the local economy without first passing through a bank as it would in other communities. These promotions are attractive: no transaction fees; no requirements for an account (keeping patrons away from other paycheck-cashing establishments and banks); plus gimmicks such as Boulder Station's "spin and win" games, the offer of a free drink for anyone who cashes a check, and the Orleans casino's "Paycheck Party Machine that allows players to pick their favorite game: poker, slots or Keno and 'Win up to $250,000.'" Casinos are such popular places for locals to cash their paychecks that reporter Adam Goldman noted that the line behind the cashier's desk at the Orleans on one Friday was longer than a football field.[2]

John Okamoto made a direct connection between gambling and the prevalence of cash in the local economy: "It [cash] is an abstraction in other cities, a symbol of power and influence. Here, it's tangible. You can go to the [casino cashier's] cage and see the money. You can handle it." Howard Schwartz said similarly that in other towns "a hundred-dollar bill is given a good look. . . . Here it is more common. Cash is king here." Even when you leave the casino, cash is still everywhere. Bernice Grand explained that one can break large bills anywhere in town, not just at the bank, and then said jokingly: "Even a kid running a lemonade stand could probably break a hundred-dollar bill."

Having heard such comments, I conducted an experiment. Around 9:30 P.M. one evening I purchased some groceries at a store near my home in Las Vegas, and requested some cash back on my debit-card transaction. I inquired what the limit was and the cashier responded, "How much do you want?" Somewhat surprised, I requested and received $250 in cash. I did the same at a store in Lawrence, Kansas, where I was told that the cap was $40 during the day and $20 at night. I have found similar restrictions elsewhere in the country. Whatever the reasons—private intentions, a hope of hiding tip earnings from the IRS, a desire to gamble with more liquid currency and in a less traceable way—cash is an important part of the way life moves in Las Vegas.

THE DOWNSIDE OF GAMBLING

Gambling has a much-touted economic benefit for locals and is a source of entertainment. Raymond Drake was quick to point out benefits he reaps because of the industry: "The gaming is what keeps me from having to pay state income tax." At the same time, Drake sees the destruction gambling can bring to a society: "The downside about Las Vegas is that if you are prone to any kind of addictive behavior—I don't care whether it is overeating or gambling—if you have a tendency for any kind of addiction, this is a place where that can thrive and take you over. It is designed to thrive on excess, and if you have a personality—whether you are biomedically prone or just have that personality—to succumb to excess, then this is not the place for you."

Gambler's Book Shop owner Howard Schwartz pointed out the dichotomy in games of chance: "I see the good and the bad side of gambling. It is a form of entertainment. It is something to do." He added that for some it can be a positive driving force in their lives. But he acknowledged it to be "a self-destructive thing too. People thinking they are going to hit the big one, and even if they do, they don't know what to do with it. . . . The local guy is dreaming of whatever he sees or hears on TV or in the newspaper about the last guy to hit the jackpot. He's like a dog with a bone who looks down and sees a bigger bone. The local gambler is on a perennial treadmill to nowhere." He gave a specific example of a longtime regular customer of his store who was involved with every kind of gambling there was before he died a decade ago. This man won $800,000 in three days playing craps. "It was a miracle," Howard related, but "a day later, he didn't have a dollar. We asked him what happened and he said, 'I'm a degenerate gambler. I was always wanting a million.' He could have taken that and no one would have ever known he was $200,000 short."

Of course, the temptation to play is always present. Tom Dennis recalled his life before giving up a twenty-five-year gambling habit in 2003: "I used to win an awful lot of money playing blackjack. Many times, probably too many, 90 or 100 times over the years, with a $200 start I'd win $2,000 or more. . . .

The next day I'd stay up all night long, get free cigarettes, free drinks, free food and wind up going broke."[3]

So much cash in the economy heightens the temptation. It is easier to spend cash than it is to get a loan, pawn a ring, or even visit an ATM. Rob Hunter of the Problem Gambling Center in Las Vegas spoke of "mixing booze with dollars inside a casino." Hunter said, referring to paycheck cashing promotions: "The idea of having an entire check in $100 bills and two free drink coupons is inherently dangerous for the problem gambler."[4]

Not surprisingly, Nevada ranks highest on the list in studies of problem and pathological gambling in the United States. The numbers from recent research are approximations and possibly underestimate the problem since "player concealment or misrepresentation," a common characteristic of pathological gamblers, is likely in phone surveys used in such investigations.[5] (For this reason researchers, such as those conducting surveys for LVCVA, often ask for budgeted numbers rather than actual amounts spent.) In addition, the numbers vary depending on the type of survey screens employed by researchers. In both measures, however, Nevada outranks every other state. Based on a screening employed by the National Opinion Research Center, Rachel Volberg reported that, in 2000, 3.0 percent of Nevadans could be considered problem gamblers and 2.1 percent pathological gamblers, the latter category being the most serious condition.[6] Those numbers compare to 1.5 percent and 1.2 percent nationally from a 1999 study.[7]

A different indicator, the South Oaks Gambling Screen, is a "primary measure of prevalence of problem and pathological gambling in the adult . . . population." From that screen Volberg concluded that 2.9 percent of Nevadans were problem gamblers and 3.5 percent were pathological in their habits. Compared to other states where this measure has been deployed, Nevada again ranks highest: the combined problem and pathological rate in Nevada is 6.4 percent, compared to a combined rate of 4.9 percent in Mississippi, the number two gambling state.[8] Based on that same 6.4 percent prevalence rate, UNLV researchers William Thompson and Kevin Schwer estimated the cumulative social costs for problem and pathological gambling in Southern Nevada to be between $459.7 and $545.2 million, depending on the cost scenario employed.[9] Additional indicators of the problem include the fact that around a hundred Gamblers Anonymous meetings are held each week in the Las Vegas area (a disproportionately large number compared to other cities) and that around 6 percent of the 11,000 homeless counted in a 2007 survey claimed that gambling was their reason for being on the street.[10]

In my research, I met a number of locals who had serious gambling addictions. Presumably for the same reasons noted by researchers above, most of my interviewees did not admit to *current* gambling problems, but in a handful of cases, I sensed that their addiction was still lurking. In chapter 3, for example, I told the stories of Charles, who had been dealt a bad hand in Las Vegas and was headed back to Salt Lake City, and Al Zanelli, who wanted to move to a city where the omnipresent slot machines wouldn't "take" all his savings. A number of other interviewees shared past experiences with gambling addiction and how they got out.

Nate Jameson, who came to the city in the 1960s with a good job, explained how he relates to the gambling scene in town. "I had a hard time with it at first," he said. He told the following story about how he got over his initial addiction: "I just walked away from it. I got tired of going broke. Even when I worked at the Mint [a downtown casino], I used to go next door to Binion's. We had a line of credit there. When I had gambled it all away, I thought, this is stupid, and I gave it up." After video poker came along, he got back into playing for a couple of years. He called his relationship to these machines, which are particularly popular with locals, "almost a hypnotic obsession." He described how, "I would go down to the Station casino, the Sunset Station (one popular neighborhood casino in Henderson), and make my football picks and would usually cash a check in. Maybe $100 or $200. I noticed that I was coming home even. See, I would take what I made on the [sports] bet and put it into the machine, too, and so I'd come home and say to my wife, 'I won again!' and she would wonder where [the money] was. I spent it on the machines. After about two years, I gave that up too. Now, I'll do twenty or forty bucks every once in a while."

Jeremy Mont didn't admit outright to having a gambling problem, but hinted at his continuing struggle to set limits for himself over the years. He grew up in Las Vegas and started gambling with fake identification when he was nineteen years old. When you live in Las Vegas, he said, "you have to have an adult mindset. If you act like a child, it will eat you up." Suggesting the impact shift-work has on the local lifestyle, he said: "I used to get off at three in the morning and I'd be wired. A couple of guys would come down and we'd get something to drink and gamble. I might make it home by 9:00 A.M. I like to drink a little and gamble, but I keep it controlled. I have friends that have drug problems, alcohol problems, gambling problems, and video poker problems. People make a ton of money and end up shoveling it back into booze or gambling." With a hint of self-reflection, Mont mentioned that he himself had partied so much at times that he didn't see the sun for three days. He said at another point in our discussion that he might lose $400 and go home and think to himself, "I didn't have $400 to lose. . . . Between nineteen and thirty-five [years old], I've probably had two to three bad nights." It is at such times that he realizes that he needs to do something: "You have to know what your limits are."

LEARNING TO STAY AWAY

Many people I met learned about setting limits at an early stage of their life in Las Vegas. Chuck Ballard told me that he is not a gambler and that you can't make it successfully in this city if you are. He learned this lesson soon after he arrived to play basketball at the university: "When I first got here I took the first check I got from the boosters to the casino and lost it in an hour [he chuckles at his irresponsible use of scholarship money]. That was all I had for the next week and a half and I needed to eat. I had to call my parents up and they sent twenty bucks to a Western Union. So, my parents sent me to college for twenty bucks. That's all they had to pay. I do very little gambling today." He said that he tells those who come to town, "If you gamble, don't move to Las Vegas."

Sandra Peters agreed. "I enjoy gambling," she told me, "but it's not something you can do here." She told me of a friend who "got sucked into it" and almost lost everything. Peters knows that personally she just can't make it a habit. Jimmy Del Toro visited Las Vegas regularly before moving here. He said that when he arrived, he had to stop gambling, except for maybe the occasional penny slots, or he wouldn't be able to make it. He then added a common refrain: if you like to gamble and live here, it will bring you down.

Several interviewees had stories of addicted and struggling acquaintances. Carlin, for example, told how his child's former babysitter lost her life savings at Palace Station, an off-Strip casino on Sahara Avenue. She has left town now, but whenever she comes back she gambles again. Carlin went on to tell of another friend in Phoenix who got a late start on his career: "I would tell him, 'Hey, just come up here and I can get you a job dealing [cards] and you can make forty or fifty [thousand] a year.' And he would tell me, 'I can't. I know I'll gamble away all my money.' Even when he visits he takes all the money he has saved up and goes until it's gone." Gambling is one of the things that attracted Carlin to Las Vegas, but he has learned to set limits for himself.

Peoria, Illinois, native Renae Shaw told a story from her family. Her sister, who had nearly completed her doctorate at Indiana University, came to Las Vegas and thought, "Whoa, look at this!" Only she eventually got caught by what Renae called the "gambling demon." "The allure here is too tempting," Renae explained. "It is on every corner. If they could confine it out of the neighborhoods and the convenience stores to just [she points to the tourist corridor] . . ." That allure was too much to handle for Renae's sister, who continues to gamble regularly today, never having completed her PhD.

Don Trimble's least favorite part of living in Las Vegas is the addiction that gambling can become for those with certain personalities. He told me of one experience: "When I was twenty-one, I was a bartender at a local bar. We had a cocktail waitress, a mother of four. . . . She was a single mom. . . ." Don interrupted himself to explain that he doesn't have a problem with tourists "laughing with their friends and gambling for entertainment." What he didn't like was "the cocktail waitress . . . who, when she is off her shift, turns around to me and hands me her hundred dollars in tips money [from the] day and says, 'Give me a roll of quarters and a whiskey.' And she turns around and plays the video poker game. . . . That part of gambling bothers me." Bob McCray said similarly: "I see a lot of self-abuse in this town. With the booze, drugs . . . and gambling. Gambling eats people up and destroys lives."

GROWING UP AROUND CASINOS

When I interviewed locals who grew up in the city, I often asked how they related to gambling after having been exposed to it from an early age. I heard a number of interesting perspectives. Tom McAllister responded this way: "I have no attraction to it. I don't understand it. It's not my juice. It is nice to be able to put twenty or fifty bucks down on a football game, but you can do that anywhere." For Don Trimble, slot machines and casinos were just part of the scenery. He told me that gambling was never a problem for him and that in both Las Vegas and Reno, where he lived before moving to Southern Nevada, he wasn't attracted to it.

Cale explained that his mom and dad never gambled, so it was never something that was a temptation for him or his siblings: "I think the environment we grow up in has an effect on us." Then he added: "I work too hard for my money to give it away. I realize that they don't build those hotels because [the house] loses all the time. And the odds aren't very good, I've been told."

Aric Walker, born and raised in the city, gave similar reasons for his lack of interest in games of chance. "Gambling never took me," he said. "I was always more interested in making money." At one point, he said, he liked to play poker and in his midtwenties he spent a lot of money on it. Then he added: "But, I had a lot. Then I just didn't do it anymore." When I asked if growing up made gambling less interesting to him, he responded, "No, not really. I have friends that are trying to make it big all the time."

Ted Burke's story of an "education" about gambling in early life focused on teachings of his parents. When he was around five years old in 1949 his mother wanted to throw Burke's brother a surprise birthday party, so his father took him and his brother out of the house so she could get things prepared. Their father drove them down to the Desert Inn, on which he had done some contract work during its construction. This was the biggest hotel in town at the time, Burke explained, and recalled the words of his father as they looked over the nearly completed structure: "These big hotels are not built with the money that people win. They are built with the money that people lose." Those words stuck in the mind of young Ted and he has never had a desire to be part of a casino or gambling culture.

Much of my own motivation for avoiding gambling stems from what I saw the first time I set foot in a casino when I was nine years old. My dad had accepted a job in Las Vegas and brought the family to town for a weekend vacation to check things out. We stayed at Circus Circus because of its reputation as a relatively family-oriented casino with a cheap buffet, the world's largest at the time, and its proximity to Wet 'n Wild (a now extinct water park) across the street. We spent one day at the water park, played games at the Circus Circus fairway, and watched circus acts directly above the casino floor. During one such performance, I glanced down to see a woman playing two slot machines simultaneously with apparent skill and prowess. Her face was emotionless and her attention fixed. That image left an impression on my young mind, which served as a personal motivation and incentive never to gamble.

Annie Abreu grew up in a house entrenched in gambling. Her dad was a dealer and casino floor manager. When asked about her personal relationship to gambling, she responded: "I've never gambled. My sister and brother haven't either. It's the only addiction we haven't had." She added that the deterrent for her was not necessarily that she grew up around it but, rather, is based on personality. "I know tons of people who grew up here that lose their paychecks all the time." Abreu noted that she is progaming and that the industry has given her opportunities she needed in her career. Based on her later comments, however, I think her father's direct involvement, and the way he taught and raised his daughters, had something to do with her avoiding gambling. She said that her dad didn't want his children, especially his girls, to get involved with the industry. (She noted that even into his seventies he struggled with women's participation in gambling.) She mentioned that her father opened the first dealers' school outside of the casinos and that she visited one time and learned how to shuffle cards. When her father found out, he was very upset. "You'd think I was on a stripper pole, he was so mad," she said.

Eloise Freeman echoed Abreu's words as she discussed her friend's parents, who had careers in the gambling business: "They would have broken the arms of their children if they had followed them into the casino [business]. All of those friends went to college and grad school. None of my friends went into gaming." Although their remarks were focused on children going into the gambling business, Abreu's and Freeman's comments also highlight the ambivalent attitude toward the industry that supported many homes in Las Vegas's earlier days.

As one might expect, especially given the comments of Abreu and Walker, gambling is a temptation for some young locals, and many underage Las Vegans gamble illegally. Rachel Volberg, in her study of adolescent gambling in Nevada, found that nearly 67 percent of survey respondents had participated in at least one form of betting in their lifetime. While that rate of participation is quite high, the expectation that it would be higher than in other states is not supported. She noted that the 49 percent of Nevada young people who gambled in the past year is lower than the "median past year gambling participation rate" of 66 percent identified in a recent analysis of various similar studies throughout North America. Furthermore, she found that the rates of at-risk and problem gambling among youth (the two most extreme stages of involvement assessed in adolescent people) are also lower in Nevada compared to what is shown in similar studies in other jurisdictions. In fact, Nevada rates were at the lower end of the prevalence range. "This is true," Volberg reported, "regardless of which problem gambling screen or method for scoring is selected."[11]

In my research, I found that the attraction exists for some young people, but many learn early on that gambling is not something they can do and survive financially. This may explain why involvement in gambling for Nevada youth is lower than in other states. Edwin Aponte, the son of a Strip entertainment producer, gave a characteristic response of local kids who might be tempted: "I think we just go through it when we're younger. Everybody has to go through that at some time; we just do it earlier." Flint Salvador told me similarly that he played around with gambling when he was a teenager, but now doesn't do it at all. Jeremy Mont, as he discussed the limits he had to set for himself, explained that it is easier to find those limits growing up around casinos, but that he still has friends who are more prone to addiction. His dad, a pit boss in the casino industry, explained that a lot of scumbags roam around in the industry, that casinos don't thrive on winners, and that he saw people throw their lives away all the time. Mont again asserted that you have to know your limits.

Russell Busch gave a more detailed explanation: "There's nothing wrong with it . . . for other people. It's not for me. I want to hold on to the money I have. But, morally, I don't see anything wrong with it." When I asked if that conclusion was easier to reach because he grew up in the city, he replied: "Every parent tells their kids that gambling is not good, but everybody has to learn for themselves. I gambled for two years and then saw how expensive it was and realized that I can't do this if I want to hang on to my money." Then he added: "I'll tell you what though, those table games were fun. Your heart is racing. They are fun! But, I can't afford to lose my money. It's kind of like a recovering alcoholic. You see it and think how nice it would be to do that again."

BUDGETS AND PERSONAL LIMITS

A high percentage of locals, as indicated by the LVCVA numbers I presented earlier, choose to gamble, but many find a way to manage it, as Jeremy Mont recommended. The experience of Bernice Grand is telling in this regard.

Sitting in a comfortable chair behind her desk, Grand looked through bright red heart-shaped glasses and told me how she came to live in Las Vegas. Several years ago she left Wisconsin to work in California. While there she made gambling trips to Nevada every three to six months. She eventually decided to move to Las Vegas in order to become a dealer herself. Unsatisfied, she eventually left her job as a dealer, but stayed in Las Vegas and still gambles. She said that if you come to Vegas, enjoy gambling, and "can make it through two years, then you will stay. You may lose a few paychecks here and there, but you'll learn to budget gambling into your life." Grand typically plays quarter slots now, but prefers video poker over the pull slot machines because she feels that video poker gives her more control. If she is winning, she will move to the dollar slots or something else that she normally would not be able to afford. "At that point," she said, "you're playing with their money and you can have a night out. It's easier when it's not your own money." She noted: "The draw, even for locals, is there because they might be the one that hits it big."

Whereas Grand's story resonates with those of other interviewees, what particularly struck me—as a nongambler—was her comment that you might lose a paycheck or two before you learn to fit gambling into your budget. Grand was one of my first interviews, and as I continued my research I found that budgeting is an important survival skill for the local gambler. The words of admonishment sung by Kenny Rogers in his popular song "The Gambler," about knowing when to hold or fold a hand of cards, seem to be true for many local gamblers.

The same day I visited with Bernice Grand, I also met Ted DeAngelo, a retiree from Brooklyn, New York. When I asked him what he thought of Las Vegas, he replied: "This is the greatest town as long as you don't gamble." At the same time he called himself a gambleholic. As we talked about gambling, even though he didn't say the word "budget," I could sense that he had developed such an attitude in the several years he has lived in the city: "The first year I was here I did OK. The second, I did fair. But from the third on I haven't done so well. I got to a point that I realized that I wasn't going to win and now I go maybe three times a week. I'll lose $60 or $80 here and there, but for me, at my age, that's OK."

I counted a dozen interviewees who testified to the importance of budgeting. When I asked Jeremy Mont whether the amount that he typically puts aside for wagering is set, he said: "Absolutely. You have your sixty bucks. You lose it and you walk away. . . . You don't go chasing the ATM for more. It's sixty bucks, win or lose." Referring to the strategy to quit while you're ahead, he added: "You have to train yourself to leave 'up,'" and concluded with the common exhortation, "They don't build these casinos off of winners."

Charley Sparks, who owns an independent barbershop in town, likes to play: "I really love the machines." But, alluding to the inconsistency in his self-employment income and the impact it might have on his hobby, he added: "You have to learn to live here. You have to know, see, that you don't have more money than the casinos." People who don't realize that, he said, will lose everything. He himself takes preemptive measures by leaving his money at his shop or at home because he knows that if he walks into a casino with $100 he may walk out with nothing: "See, I could have $95 dollars in a machine [in credits] and I might think, 'I'll just get that last $5,' but I could end up with nothing in the end. . . . You just have to know that if you have the $100 or $20 or $5 in nickels, that is what you will spend."

Retired schoolteacher Shari Nakae described her method of budgeting. She said that when she first came to town, a group of teachers would go and spend $25 on payday: "I would play blackjack. So, that was $25 a month. After we had [our first child], I would go once a year and spend the same amount as the number of years I was at that time. So, when I turned thirty-one, I spent $31." These days, she might spend a total of $60 in bets for the whole year. She said that she always has a set amount when she does play.

Interestingly, many of the budgeters were the same people that decried the pitfall that gambling can be. Howard Schwartz said that he plays only maybe $10 a week, bets on horse races throughout the year, and on occasion, "I play poker, if I'm lucky." Carlin explained that since he deals cards for a living, he doesn't do much of that: "After forty hours a week of it, I don't want to get close to the tables." But, as a sports fan he likes to bet at the sportsbooks on occasion. He said that he does well at it, having worked previously in that sector of the casino. He knows what he can do financially: "I just have a budget allocated." Sandra Peters and Renae Shaw have a similar philosophy and, like Nakae, gamble only with visiting relatives or friends. Shaw, for example, told me that sometimes she will gamble when she goes out with her family, but asserted that she will take only $10 and no more and that it is "more of a game" at that point. She likes to play the penny slots. She added sarcastically: "I don't understand the addiction thing. It's really quite boring. How many different cherries can you see?"

You will recall that Bob McCray doesn't like the dark side of gambling, but admitted that he gambles at times, typically only when friends come to town. He claimed self-control and said that, if he goes, he will do only $200. However, he confessed that sometimes he will splurge and do a thousand or more. He recalled one experience when having a declared budget was insufficient protection. A while back he and his girlfriend went to the Palms and lost everything in his pocket, something like $6,000. "They get you with the booze," he lamented. "They keep bringing it out and pretty soon you're . . . [he makes a gesture signaling that he doesn't know what he's doing at that point]." For that reason, he concluded, "I don't go out that often."

When I asked Ernest Bannister how his life in Las Vegas is different from his former home in upstate New York, he said: "It's about the same. There's nothing that we really miss." Then he spoke of the uniqueness of plentiful gambling opportunities: "Gambling is available, of course, in Vegas. I like the local casinos—the casinos that cater to the local population. My wife loves bingo. I don't care for gambling, but I provide all the gambling money for her."

We have an amount set aside for that. We knew coming here that we would have an entertainment budget, and we stay with it pretty much."

The entertainment value in gambling of which Bannister spoke was mentioned by others as well. Jeff Simmons noted that he bets on sports quite often, but also likes to play video poker on occasion for fun: "The other day I went into a Terribles [a local gas station/convenience store chain] and put ten bucks into a machine and played until it was gone and I left." Aric Walker, who said that "gambling never took me," added later in our conversation that he will go to the casino on occasion because his wife likes to play a video poker machine called Triple-Play Deuces. "It's entertainment for us," he said and returned to his criticism of the betting habits of some: "It becomes so much more for other people. People come here thinking they're going to clean the town, but that's not what's going on down there."

Whether the local gambler's desire is for riches or just plain fun, based on what I heard and saw in the city, it comes down to the balance between limits and desire. I like Jeremy Mont's remarks about the importance of one's own choices: "You have to take personal responsibility for your [stuff]. It's just like anywhere else. There's no beach. We have casinos. If you go to the beach, you might get bit by a shark. If you can't swim, you don't go into the water." I like even more Howard Schwartz's admonition for moderation using a different beach metaphor: "Gambling should be a thimble full of sand on the beach of life. Of all the sand on that beach, gambling should only be this much [he makes a small circle with his forefinger and thumb]."

A SEXUALLY EXPLICIT TOWN

The local connection to the tourist Las Vegas is also plainly manifest in the sex industry. I have already discussed several ways in which locals can succumb to the various vices in the city (see chapter 3), but here I want to talk specifically about feelings toward the landscape of sex openly visible in both tourist *and* local sides of the city.

To clarify, prostitution is illegal in Las Vegas, contrary to what many outsiders believe. In fact, legal brothels can be found in only ten of Nevada's rural counties. Of course, despite the law, prostitution is nonetheless widespread in Sin City. As one indication of the proliferation, consider what a longtime police officer on Metro's vice squad explained about his division's work and reputation: "We used to go to LAPD to get training. Now we're setting the bar."

A number of interviewees commented that life can be a challenge in a city where a woman's body is often (some say overly) objectified. When asked whether or not the glamour of the Strip carried over into local life, Joseph Mani responded: "Very much. It's natural. You can't live in a fish bowl and not be affected by it." He then noted that single women have a lot to compete with because of the city's sex industry. Mavie Roberts, who loves the hospitality found in the city, pointed out that, just as a higher level of service is expected, women in Las Vegas have to live up to a higher standard: "Everywhere you look there are naked women, and you see what guys want and what guys expect." Roberts, a single woman in her twenties, felt that this part of the city's culture pressures women to meet such expectations.

Sandra Peters was more direct in describing the difficulties she has experienced as an attractive single mom in Las Vegas: "I'll take my daughter down to the Strip sometimes. . . . It's fun to see the Bellagio fountains and stuff. . . . I took her into the Wynn with me to pick up some things from a client and these two guys out front were picking up on me . . . and I was with my daughter!" She continued: "It's because of what Las Vegas is. It's where you can get rip-roarin' . . . drunk," and where guys can go to the strip clubs and pick up a girl.

In her essay "Who Puts the 'Sin' in 'Sin City' Stories?" sociologist Kathryn Hausbeck described a similar experience. When she initially came to Las Vegas to interview for a faculty position at UNLV, she got sick in the middle of the night and ventured out of her hotel and across the street to a convenience store to purchase some medicine. In "shorts and a tank top," her "blonde hair swept into a messy ponytail," Hausbeck heard catcalls along the way and endured staring eyes as she entered the store. After several moments, she understood: "They all thought I was a prostitute." She was formally propositioned outside after she left the store and wondered: "How had I, a young but serious grad student, whose biggest concern was presenting herself as a legitimate academic at the job interview, come here and ended up with a flu that transformed me in the eyes of the streets and the city into an entirely different kind of working girl?" Within three months of taking the job and moving to the city, Hausbeck was propositioned two more times while she sat at a stoplight in her red car. At that point she decided to focus her future academic research on the sex industry. She encapsulated the position of some women in this environment in the following statement, made after she had been propositioned the second time: "The equation was as simple as 1-2-3: I must be for sale. It began to make sense, and I began to make sense of myself in this city."[12]

Many locals strive to remain separated from what Peters called the "seedy side" of the city. As we concluded our discussion she noted: "I don't expose myself to that part of it [the city]." After giving a general description of life in Las Vegas, Bonnie Pratt told me: "I know the other is there, but we're not affected by that." I asked her what she meant by "other," and she replied: "When I first came here and opened the newspaper, I couldn't believe the kind of things they were advertising. The pictures totally shocked me. . . . The gambling, the promiscuity, the drinking . . . we've made it so that doesn't affect us."

The sex industry in Las Vegas, however, is impossible to completely avoid. Locals see it every day, for example, as they commute on valley roads. Such is particularly the case two or three miles to the east or west of Las Vegas Boulevard's transect through the valley. Strip clubs are typically in plain sight of several main traffic arteries through the area. Elsewhere the presence of the sex industry is announced on billboards, taxicab displays, and roadside newsracks containing leaflets advertising escort services and the like.

The shortest route on my twice-weekly trip from the east side of the valley, where I lived, to the west side, where my five-year-old son's prekindergarten sports class was held, was Desert Inn Road, a street notorious for risqué billboards. As we passed through that particular stretch, I made a habit of asking Caleb random questions to draw his eyes and attention away from signage that revealed more of a woman's body than I would wish him to see. I was not alone in my concern. Mother of four and billboard protester Shari Peterson had a similar experience: "As I drive with my young children along our streets and highways, we are forced to observe and subject ourselves to billboards with sexually explicit and revealing content. My family has lived in Las Vegas for more than 10 years, and over those years the adult entertainment billboards have become less and less interested in keeping at bay the sexual content of the businesses being advertised."[13]

Debates over the landscape of sex in Las Vegas often enter the public discourse. The comment from Shari Peterson comes from a news article that presented representative voices in a debate about whether or not suggestive and sexually revealing billboards should be more strictly curtailed. One side says, as Peterson did, "Our children have the absolute right to be protected from viewing adult entertainment billboards, including advertisements on taxis, that are not suitable for the general public." The other, commonly voiced by representatives of the American Civil Liberties Union, says in rebuttal: "the calls for censorship of those [ads] containing sexual innuendo should raise concerns among those who take free speech seriously"; or "the ads in question are not illegal. They aren't obscene under Nevada law, nor do they fit within Nevada's legal definition of material that is 'harmful to minors.'"[14]

Other debates have surfaced in recent years that further highlight the contentious atmosphere in Sin City. Even though prostitution is not legal in Clark County, arguments about that industry surface every now and again. Such was the case in 2007. First, a federal judge struck down a state law that prohibited legal brothels from advertising in the off-limits Nevada counties. This prompted a public discussion and an appeal by the state's attorney general. Next, a vehement dialogue erupted in the pages of local newspapers—including at least six letters to the editor and four entries by regular columnists—when Mayor Oscar Goodman told a *New York Times* reporter (as he has done on other occasions) that the city should be discussing legalized prostitution.[15]

In the fall of 2007 the Las Vegas City Council debated for nearly a month and a half whether or not to grant redevelopment money to Olympic Garden Topless Cabaret for remodeling its frontage along North Las Vegas Boulevard. The project itself conformed to redevelopment stipulations, but concern existed over the images that would be displayed on the marquee. The council finally pledged $50,000 for the improvements, but not until Olympic Garden agreed to scale down the video screen on which revealing images would be displayed to passersby.[16]

In 2010, the Clark County Commission took up a similar debate regarding another local strip club's promotional campaign. The "stripper mobile" is a modified cargo truck on which the box has been converted to a plexiglass-enclosed lighted stage (complete with stripper pole) wherein scantily clad dancers perform to advertise the main establishment as the vehicle traverses Las Vegas Boulevard and surrounding thoroughfares. Citing safety (how this might distract gawkers driving past), legal (using public rights-of-way for private advertisement is prohibited), and decency concerns (including one grandmother who complained that she and her grandson saw the "advertisement" pass them along Paradise Road, near the strip), county officials hoped to stop this new style of mobile billboard before it became a trend.[17]

AMBIVALENT ATTACHMENT

The mere existence of such community debates over topless cabarets and adult-oriented advertisement in Sin City might surprise and intrigue an outside observer. After all, as urban planning students learn, the acceptability of explicit forms of advertising is generally dependent on the community's culture of tolerance for such businesses and their promotional strategies. Stated differently, one might expect that since Las Vegas is known to embrace sin, the community would not oppose an overt presence of sexual images in the public sphere. The public debates over sexual expression are intriguing because they illustrate the ambivalent acceptance of the sex industry in Las Vegas. Even though the presence of the sex industry is a part of life in this city, many locals choose to believe that it isn't or shouldn't be part of *their* particular lives.

Sensitivity among Las Vegans about sexually related issues often boils down to the bifurcation of the city into its tourist and local halves. Columnist Jon Ralston refers to this in his response to Mayor Goodman's comments to the *Times* reporter about prostitution: "[An] argument worth having . . . is what kind of community we strive to be and whether we are content to tell a tale of two cities. One is where Goodman, a Dickensian character if ever there were one, sets the tone, a place where anything that's legal is just fine, where taste is optional, where no sin is too sinful. The other is a wholly different venue, one where parents are proud to raise their children, where culture, sophistication and erudition are prized, where family values refer to something other than mob mores."[18] That same sentiment was voiced in arguments for and against the Olympic Garden sign, revealing billboards, and stripper mobile issues.[19]

The ambivalence expressed in "it's okay for the tourists as long as it doesn't affect me" appeared in a number of interviews. Several people commented on the practice of canvassing walkers in the tourist corridor to advertise escort services, strip clubs, and so on, an activity that has caused its own set of debates in the public sphere. As we discussed aspects of raising a family in Las Vegas, Don Trimble, who doesn't have children, said: "I'm not sure I could raise my kids here because I don't want them to see the billboards with strippers on them. Now, I don't mind the selling of Sin City, but I was walking with my mom on the Strip a couple of years ago (we were out just to have some fun away from the house) and I was so embarrassed when those people on the Strip were handing out leaflets." Sandra Peters said similarly: "I don't like [going to the Strip] because of the people on the corners handing out the leaflets. I don't have a problem with what they are doing, but you'd think there'd be a little more respect for mother and daughter." Another interviewee noted simply that he doesn't like the nonconsensual aspect of the unavoidable promotional material.

John L. Smith doesn't take issue with sexually oriented businesses and their place in the city's culture in principle. At the same time, he identified the sex industry and its trappings as one element of the city he wished didn't have as much influence: "Las Vegas is really good at marketing sex and that makes it hard on people and families. It's hard on women. It's hard on men. It's almost like it has a coarsening effect on people. It's part of the tourist culture that has leaked to the local. There are shiny, Playboy-quality magazines that you can get for free [on newsracks around town], all naked. They spend a fortune on this stuff. It's almost getting to a point where it has an Amsterdam quality. . . . It [the sex industry] has the most potential to sink us in terms of community stability."

Locals' distaste for the city's overt sexuality is not limited to the selling of sex in the tourist corridor through billboards, leaflets, or other publications. For evidence consider a seemingly benign but illustrative example. In a Sunday Week in Review section of the *Review-Journal*, editors placed a picture—unrelated to any news item in the column—of a man with his hand on a woman's breast with the caption: "A feel for the moment: Kyle Stocking grabs his wife, Francie, on Saturday at Las Vegas Motor Speedway during NASCAR Weekend events."[20] That image prompted two immediate and scathing responses in the bloglike comment section of the article as well as three critical letters to the editor published a few days later.

Here are some of the remarks. Pat Bisceglia wrote: "Don't you think that the picture . . . was over the top. There is a reasonable possibility that children will see it and will get a distorted view of accepted public behavior." Jenna complained: "I am offended and disgusted by this crude . . . photo. Not only is [it] totally inappropriate to run this . . . in the first place, but it's [*sic*] vulgarity is amplified by the sleazy caption that you've added to it."[21] Lisa McKenzie wondered: "What is the purpose of taking the picture, let alone publishing it? Show some class, for God's sake." Carol Vick was shocked: "Of all the news that is out and about, stories waiting to be told, you put that in? What were you thinking?" Mary Ashcraft wondered the same and added: "I am surely not the only *Review-Journal* reader who is offended. That is most certainly a new low for the *Review-Journal*."[22]

Now contrast that rejoinder with the lack of reaction to a similar photo that ran in a story about "artistic diversions" for tourists on the Strip two years earlier. One piece of art discussed therein was a bronze statue of the near-fully exposed backsides of the Crazy Girls topless revue kick line, the same one pictured in the famous NO "IFS" "ANDS" OR ("BUTTS") billboard and taxicab ads local parents have often complained about. The accompanying photograph is of a male tourist grasping the buttocks of one of the bronzed figures.[23] No complaining letters were published in response. I doubt anyone even noticed given the more revealing advertisements they face on a daily commute to work. Understandably, the cases are different, but it is that difference that makes the point that "it is okay there, for the tourist, but not in my Las Vegas, not for the local."

In a similar vein, Las Vegas residents have often complained about revealing or risqué art at the Las Vegas Art Museum, the city's municipal art house once located miles from the Strip on West Sahara Avenue before it closed amid financial difficulties. Libby Lumpkin, the museum's former director, said in an open house lecture that she has been criticized for the nudes shown in some of the art displayed in the gallery and received a "strongly worded letter to the museum" in response to a Kaz Oshiro piece depicting "666" inside a Christian fish symbol.

For many locals, such views are simply hypocritical. "You live in Las Vegas, deal with it," the argument goes. Although she was speaking about gambling, Shawn Newman's comments apply equally well here. She remarked: "I hate people who say they hate gambling. I may not like it, but I appreciate it for what it is. . . . But people who say they hate gambling and live here is like when people say they hate Mormons and live in Salt Lake. I just want to tell them they can live anywhere. If they want to live here, deal with it." Similarly Annie Abreu criticized a person who complained about a professional meeting held in a casino. She exclaimed what she would like to tell such a person: "Shut up! . . . You live in Las Vegas!"

Hypocrisy may exist in this overlap between tourist and locals place, but I don't think it's so simple. On the one hand, locals cannot just ignore the sex industry in their city or make it go away through activism and protesting; it is part of Vegas life. So they should open a dialogue about the future of Las Vegas that would entail what locals want their lifestyle to be in terms of sex, gambling, and sin. Or perhaps locals just need a break from their embedded interaction with the city's landscape of sex. Consider art museum director Libby Lumpkin's sympathetic and insightful assessment of why some Las Vegans struggle with racy art. She told me that the local sensitivity to such things comes from being surrounded by sex and vice every day; they "look to culture as a respite from that."

RAISING CHILDREN IN "VEGAS"

In a city whose engine runs on the fuel of gaming and adult entertainment, raising a family is an obvious concern for many locals. When I asked current and potential parents what they thought about raising children in Las Vegas, the responses revealed another important aspect of the local relationship to the "Vegas" image.

In the opinion of some, the city's twenty-four-hour culture greatly impacts family life. As an example, I'll describe the typical day of one Las Vegas couple as explained by Rio Garcia. Garcia works as a security guard at one of the large resorts near the Strip. His wife is an assistant manager in retail at MGM Grand. His shift goes from 4:00 P.M. to midnight. She works from 5:00 A.M. to 1:00 P.M. Garcia comes home and goes to sleep at about 1:00 A.M., then gets up at 4:30 to see his wife off and walk her out to her vehicle. He sleeps again until around 9:30 or 10:00, when he gets up with his twin sons. He feeds them and "does the climb on daddy thing and then whatever." Garcia said the boys eventually get restless and need to get out of the house, and so every once in a while they go to the MGM to get mom as she gets off work. If she has to work late, they visit the casino's lion habitat or go to the kid-friendly Rainforest Cafe. Such is the life of a Las Vegas couple trying to make it.

The impact erratic schedules can have on families is undeniable. One marriage and family therapist explained how the "24-hour way of life is probably the single most destructive element for families." He noted: "Shift work creates a lot of problems with relationships. . . . Many kids don't have much supervision, and no extended family to watch out for them if their parents work."[24] When I was growing up in the city, many of my school acquaintances were what I later heard identified as "latchkey kids," returning from school to an empty home. One interviewee, Andre, noted that it is easy to find work in Las Vegas, but parents need to wonder who is going to be influencing the children while they are out of the house.

Judy Haynes recognized such difficulties. As my children played with her grandchildren in the park one summer afternoon, she explained her wish that she could have raised her children in a small-town environment. In fact, she lost one daughter to drugs and mental illness and is trying to raise two grandsons from that daughter. Haynes drives the Deuce—a double-decker bus—on Las Vegas Boulevard starting at 2:00 A.M. each day. This is necessary to provide for her family but affects her ability to raise the children effectively. She spends afternoons with the kids and tries to sleep in the evenings and late mornings after she gets off work. She has renters in one of her rooms and they help take care of the boys, but, she lamented, the woman smokes marijuana and sleeps a lot during the day when she is supposed to be helping with the boys.

A number of interviewees expressed concern about the lack of activities appropriate for children in an adult town. Renae Shaw noted: "I know I can take them to the park. I can take them to the museums, but how many museums do we have? What can you do with little bitty kids? What do nineteen- or twenty-year-olds do? Nothing. Go to jail?" That last comment made me wonder if that was the story of one of her children. Throughout the interview her comments suggested that something like that had scarred her. Her answer to my question, What is the one thing you would tell an outsider? is revealing: "For me, [I'd say] 'Don't move here.'" She noted that it depends on the reasons potential newcomers have for relocating to the city, but, "If they are coming to seek a better life or quality of life, then don't come here. Go to Utah or some small city somewhere." She noted that her niece had just moved from Las Vegas to Wichita because she didn't want to raise a family in the city where she grew up.

As one might expect, several interviewees spoke specifically about temptation's effect on their ability to raise children. Carlin said that his perception of Las Vegas has changed with life's stages: "Before, when I was single and then childless and married, it was okay. Now with a kid, I'm wondering if it's the best thing." When we met a day later in his home, as his four-year-old napped on the couch, I followed up on that thought. He said that he doesn't want his son to have the exposure to the stuff you see on the billboards and such. Orthodox rabbi Levi Behar explained that his least favorite part of living in Las Vegas was the billboards. I asked if that was both as a father and rabbi. He responded: "Yes. But, more as a father than a rabbi." He noted that the prolific nature of sexual imagery is even more damaging for his children than gambling because it is more "in your face."

Jake Glennon raised his children in the city, but looking back on it, wondered about its effects: "We've enjoyed Las Vegas the full time we've lived here. There is always something to do. And on the business side, Las Vegas has been very good to me and my family. But, looking back, I probably would have raised my kids in another town. I would like not to have them exposed to things in the casino industry. That's not to say that my kids didn't turn out all right in the end, but to find out that at sixteen they had fake IDs and were hanging out in bars and clubs . . . I mean I did it in Colorado, but here there's so much more. They are all good kids, so I can't complain, but . . . if I had my druthers, I would probably have raised my kids elsewhere."

One pastor of a nondenominational church moved to Las Vegas in the early 1990s. His children were ten, six, and two years old. When asked how he felt about raising kids in the city, he told the following story: "My senior pastor's prophecy was, 'You're going to lose someone to the Strip.' That's how he said it. When your home is solid, you have your own values and your own safety and security in any environment. But, the environment does take its toll." He didn't say exactly what happened with his kids, but he commented: "The familiarity . . . with some of the sins that are here that have an overall impact on them hasn't been devastating, but it does have an impact."

The pastor's comment that the home can provide security reflects the view of the majority of people I talked to about the topic. That is, parenting is a choice and a parent is responsible for teaching and communicating with children about the challenges they'll face, in whatever place and whatever environment. In the view of many observers, an active and involved parent has a high likelihood of being a successful one in Las Vegas.

Consider also Flint Salvador's response to my question about whether more temptation exists for kids in Las Vegas: "Well, the religious answer is no. Everyone is tested equally; there are just as many choices to be made no matter where you are. The nonreligious answer is yes. How could there not be? But, really it's the same, just different choices." Salvador compared the experience of a teenager in Las Vegas to one living in rural Nevada, where he had spent time in younger years: "At night, when [Las Vegas] teenagers are looking for something to do, they have more choices. In Alamo [Nevada] they might just be able to try and get some beer or go to Ash Springs or steal farmer Bob's cows. But in Las Vegas they might try to sneak into a cathouse or to a beach at the casino or to gamble or something. It's not that one place is better for kids than another. There is just more to get into in Las Vegas."

Tina Lewis spoke about the need for proactive parenting. She pointed out the opportunities to keep kids busy in constructive activities: "Parents have the same avenues as in other communities; you just have to find them." Then she addressed some of the challenges unique to the city: "We have to educate our kids regarding gambling and drinking," specifically, she said, in determining for themselves the line between use and abuse, between doing it for fun and doing it in excess.

Aric Walker, who was born and raised in Southern Nevada and is now raising three kids of his own, encapsulated the philosophy of an active parent: "It all comes down to parent involvement. We like to be involved with our kids." Then he referenced the difficulties created by the town's twenty-four-hour nature: "Parents that have problems are those who don't have the time to spend being involved with their kids." He noted how his own mother and father were involved in his schooling, while many of his friends who didn't get that involvement "didn't turn out so well." He continued: "A lot of them are dead actually." He feels that teaching values, discipline, and the importance of work are essential solutions to the problems that Las Vegas parents face. Summing up, he said: "It doesn't matter if you're in the middle of Sin City like we are"; you can still raise a good family.

Even some of the parents concerned about the city's impact on children agreed that proactive parenting is a key to keeping your child out of trouble in a city where the traps and temptations are plentiful. Tom McAllister, for example, explained the one thing he would like to tell an outsider about his town: "Ten years ago, I would have said it's just like any other town. I grew up as an alter boy, in cub scouts, and little league. It's not a family place [today]. It's a modern day Sodom and Gomorrah. It is Sin City. The word 'No' doesn't exist in Las Vegas." When I asked him about raising his kids in that type of city, he responded in a slightly different tone. "It's hard," he said, noting that he dislikes the images his kids see at such a young age, but then added: "Parenting is tough in any town, at any age, and in any environment." In the end, parenting is a personal thing, and you deal with the circumstances given you, including the place in which you raise your children.

The following stories illustrate well the extremes of parental involvement and the foreseeable consequences. Raymond Drake told me of a recent visit he made to a neighborhood drugstore. While there he watched a man and a woman with kids in tow approach the store. Then he noticed the mother and father sharing a forty-ounce beer along the way, finishing it off before entering the store. He watched them play the slot machines for a time and purchase another "forty" to drink on the way home. Raymond exclaimed wryly: "Now let's talk about [parental] modeling."

Contrast that with Trish Allison's experience as a single mom of two boys. She acknowledged the difficult environment: "The first billboard my son read was 'totally nude.'" When asked how she would describe the place to an outsider, she modeled the proactive parent advocated by so many interviewees: "I tell them, 'Don't believe the stories you hear.' I think Vegas gets a bum rap. Okay, so I have two boys in a sexually explicit city. I think it makes it a conscious topic." She said that she has sat them down and talked to them about respect, both for women and for themselves. She encourages them to call her if they think they're getting into a car with a drunk person. She has talked with them about "after parties" some schoolkids are holding after formal dances. Allison is getting at the notion of openness in the community I brought up earlier, especially as it relates to the sexual explicitness of the city's culture.

Like Trish Allison, Ted Burke sees the positive angles and opportunities for parents in Las Vegas. When asked for his thoughts on raising children, he responded optimistically: "There's so much more positive here than any other place. How can I say this? It's hard to [he paused] . . . Positive values can be taught in a negative environment only if you have an equally strong positive environment." Burke spoke of opportunities that he and his wife had to move to Phoenix, Reno, or Sacramento. In the end, after they seriously looked at the options, they decided to stay because of a greater opportunity for good in Las Vegas. Whereas the choices for kids in some communities may fall in a gray area between what their parents say and what the community accepts, in Las Vegas it seems that same choice is the starker (and possibly simpler) one between black and white.

A SYMBOLIC LANDSCAPE

Geographer Donald Meinig has written a classic essay on idealizations of American communities. In it he proposed: "Every nation has its symbolic landscapes. They are part of the iconography of nationhood, part of the shared set of ideas and memories and feelings which bind a people together." Such landscapes portray "images widely employed because they are assumed to convey certain meanings." Although the most obvious examples in this vein include the White House and Independence Hall, Meinig presented three others—the New England Village, American Main Street, and California Suburbia—as national landscapes that portray "a particular kind of place rather than a precise building or locality." Further, he noted that in using such examples, the goal of landscape observers is not to determine that a landscape speaks to a certain cultural attribute, but to determine what it "says" to us in terms of what it represents.[25]

Even though Meinig's focus is on national symbols, his work can apply to other scales, such as the city. The twenty-four-hour gambling and entertainment landscape of Las Vegas is one such symbol for this city. More

Galleria Drive and Patrick Lane near the exit off Highway 95 to Russell Road. Aside from a handful of multistory office, government, or condo buildings scattered across the valley, casino-hotel towers are the only buildings that rise high above the valley floor. This view of Sunset Station in the city's southeast corner represents a striking contrast between suburban locals casinos and the sprawl of one- to three-story buildings that carpet the valley floor. Photo by author, May 2008.

broadly speaking, the Vegas image becomes a symbolic landscape not only literally but figuratively in that it influences and represents daily local life in sometimes unnoticed ways.

In making such an argument, I would first point to how that image and perception have become infused into local culture. In this chapter I have discussed, for example, the "cash-town" mentality that springs from the local gambling landscape. In previous chapters I presented the benefits locals enjoy in terms of lower taxes and a vibrant economy; the self-centered nature of Las Vegans that springs from the gambler's interest in his own winning; the culture of newness that mimics a trend to build bigger, better, and newer casinos for an ever-changing clientele; the pitfalls into addiction and despair that can develop from involvement in gambling; and the intrusion of Vegas-only quirks into daily life.

Second, I would point back to the viewshed map I presented earlier to illustrate the ever-present nature of the Strip landscape. That such a prominent and iconic display of the gambling and entertainment industry is so widely seen by locals on a daily basis is likely to influence their attachment to a place that is inseparably connected to that same industry. Additional evidence is that the four main east-west cross streets to Las Vegas Boulevard are named after Strip casinos built during the postwar boom: Tropicana Avenue, Flamingo Road, Desert Inn Road, and Sahara Avenue. Each of these arteries stretches across nearly the entire valley and is traversed by tens of thousands of people every day. Not only this, but many businesses have taken their names from the streets they line and, subconsciously, from the casinos that gave their names to the streets in the first place. In short, the tourist landscape of the Strip is not only visually ubiquitous but has left an indelible impression on the interconnected local landscape.

Finally, I would draw attention to the pervasive presence of legalized wagering and sexualized imagery in nearly every aspect of life in Las Vegas that I have presented in the pages of this chapter. Local residents see this wherever they go and many participate directly in it. For some, it is part of their entertainment lifestyle. For others, a line of video poker machines or revealing billboards may be a mere backdrop to the mundane experience of driving across town or paying for gasoline at a convenience store. For both, gambling and sex are simply there, so much a part of city life that people talk about them as if they were as normal as church steeples in a New England town or freeways in Southern California.

I will conclude with the words of two locals, one an occasional gambler and one not, that aptly describe local views regarding the Vegas image. Their comments also illustrate the uniqueness of the gaming landscape in Las Vegas and the normalcy that it becomes for the local. When asked if he gambles, Darren Sedillo responded: "Yeah, I gamble once in a while. When we visited [before we moved here] we used to look forward to the gambling, but now that we live here, it's almost like we don't even notice it." He added that going into the grocery store today, you don't even notice the gambling or the sounds; it is so much a part of the everyday landscape.

Longtime local Eloise Freeman described her relationship to the gambling landscape similarly: "It is just as ordinary as anything. At the grocery store, there is the bread department and the gambling department. It's that normal. Now, I don't visit the gambling department. Because of what I do [for work], I go to the casinos a lot. When I do it's like how in your living room you might have a sofa in the middle of the room in front of the fireplace. You don't walk over it when you go through the room. You go around it. When I go into the casinos, I just walk around that part of it. It is like wallpaper. It is part of the scenery."

EIGHT

Religion in Sin City

eligion may be the last thing that comes to mind when people think about Las Vegas. Jud Wilhite, a pastor at a large local nondenominational church, recalled his attempt to convince a Virginia woman that churches actually do exist in the city. After he told her what he does for a living, she replied confidently: "No, you aren't. There are *no* churches in Las Vegas." According to Wilhite, "Her certainty was absolute. . . . Her perception of the church just could not make room for Vegas."[1] Other clergymen echo this experience. Local Imam Quadir Nassif said: "When we talk to Muslims around the country, they don't believe us that there are Muslims in Las Vegas." Orthodox priest Father Kent Sharp, Catholic priest Father Frank Green, and Jewish rabbi Josef Rothschild each has been asked if they have slot machines in their parish halls or synagogues. The stereotype is strong enough to prompt countering jokes. Father Green offered: "What happens in Vegas, stays in Vegas, especially if you go to confession," while Rabbi Rothschild provided the more sarcastic: "Yeah, if you get a Torah, Torah, Torah, then the Ark opens up."

In reality, faith is an important part of life for many residents in this town known for sin. While all the religious leaders I spoke with acknowledged the Sin City image, they also explained that the religious community was large, diverse, and growing. Some, like Orthodox rabbi Levi Behar, treat the paradox with humor. He told me how he responds when people ask what it's like to live in Southern Nevada: "I tell them that Las Vegas is the most religious place on earth. There are more people praying here than anywhere in the world, more than in Jerusalem. And here, when you're at a table and need a certain number or a certain hand, you really mean it." In a more serious tone, Dr. Melvin Roberts, who leads a Baptist church in West Las Vegas, eloquently stated: "In Las Vegas, God is not dead. From the Latter-day Saints to the Pentecostals to the Baptists to the independent interdenominational churches, God is alive and well in the worship life of his people." In this chapter I explore what this worship life looks like.

As with each sense-of-place theme I have presented so far, religion has its own local cliché: "There are more churches per capita in Las Vegas than in any city in the country." Such an assertion is probably just another Las Vegas myth, especially given the city's recent growth. Furthermore, the numbers—from published relocation guides or national religious censuses—that could be used to verify it are suspect. But, true or not, the saying suggests something about the personality of this place. Perhaps the church myth is necessary for locals to, as Catholic bishop Daniel Walsh put it, "counter the image . . . projected in the media that (Las Vegas) is Sin City." (Such a view conforms with the more general assertion discussed in earlier chapters that some Las Vegans see their city as a normal place.) According to Frank Beckwith—a Baylor University professor of philosophy and church studies who grew up in Las Vegas—the myth may have originated in the need of Las Vegans to feel secure with the popular image of their place: "Perhaps [Las Vegas] needs more forgiveness, so there are more churches."[2]

More than anything, however, the claim of so many churches illustrates the dual nature of Las Vegas that is my central thesis. The worship life in Las Vegas is comparable to that of any other city in the country in some ways, but in other ways it is different. I will focus mainly on the uniqueness. By understanding that experience, I hope to highlight more clues to discovering the sense of local place in this unique city.

In order to study the influence of belief systems on the personality of "Sin City" and the impact of the city's culture on the lives of believers, I interviewed twenty-seven members of the clergy across the spectrum of faith traditions in Southern Nevada, using a sampling strategy to gain access to viewpoints from each type of creed. Following the lead of other religious scholars, I classified faith traditions represented in the region into major groups—based on social characteristics of belief systems and how they relate to the secular world of adherents—and selected interviewees from each group. The eleven classes I used include Roman Catholic, Evangelical Protestant, Mainline Protestant, Nondenominational, Black Protestant, Liberal (Unitarian Universalist), Mormon (Latter-day Saints), Confessional Protestant, Judaism, Muslim, and Eastern Orthodox/Old Catholic Church. I interviewed at least two representatives in each of the groups, with the exception of the liberal Unitarian Universalists, who have only a single congregation in Las Vegas.

As the head of a congregation, the cleric represents the belief system of a group of people. As the guiding figure of their faith's followers in the area, they are aware of the teachings and doctrine of their religion and how those might converge and/or conflict with the environs of Las Vegas. At the same time, religious leaders are themselves believers, practitioners, and insiders to their faith. Furthermore, clerics typically dedicate much of their lives to meeting the needs of congregants and thereby can potentially understand the collective experience of a larger set of believers. From a practical standpoint, then, one interview can (and did) provide insights into many lives. Whereas the perspective of clergy can result in a partial view of religion in the city (i.e., the intent and ideals of the leader), the head of a congregation was able to reflect on and describe those attitudes I was interested in gathering without divulging sensitive or private information of the congregants.

A MODERN-DAY CORINTH

As one might expect, several religious leaders more than merely mentioned the Sin City image; they focused on it. A number of Protestant pastors evoked the image of the Christian apostle Paul's mission to the Corinthians. Jud Wilhite explained how Corinth, like Las Vegas, was a tourist town, a service-based economy, and a place known for immorality.[3] Dr. Hank Taylor (who likes to go by Dr. Hank) didn't hesitate to draw the comparison: "Corinth was the Las Vegas of the Western and Eastern worlds. It was Sin City." One of the most quoted New Testament scriptures, Paul's sermon on charity in 1 Corinthians, chapter 13, he noted, "was written to an audience of the first-century Sin City." He continued: "Can any good come from Sin City? Absolutely. In fact, I would wager [he chuckled at the pun] that there is a lot of good that can come from Sin City. I think people of faith recognize a great need here."

Another Lutheran pastor, Ian Sears, described his first thoughts on receiving the assignment to Las Vegas: "Prior to coming here, I thought what a lot of people do . . . gambling, the party place, and prostitution. . . . [Then] it came to mind about Paul going to Corinth. Corinth was similar to Las Vegas. It had a lot of sin. But Paul went there and set up a church. People were there who believed. There are a lot of people here who believe the word of God. People come here for different reasons. . . . But, they also want to worship." He explained the answer he received after asking God, "What are you doing here?" "I realized that I had an opportunity to make a difference in a place where people really needed God. What better place to go to proclaim the word of God? What better place than the place of sin? Maybe we can make a small dent in that. Maybe we can make a difference." Sears pointed specifically to the adult entertainment industry: "There is a constant opportunity for ministry. You just need to find a way in."

Some groups have found their way in. Former call girl and prostitute Annie Lobert started Hookers for Jesus, a nonprofit organization and ministry to support sex workers who are seeking a new life. In fact, cable's Discovery Channel ran a short *Investigation Discovery* series in 2010 profiling Annie and the lives she's affected in "Hookers: Saved on the Strip."[4] Similarly, XXXChurch.com, a national nonprofit set up by the anti-porn pastor Craig Gross, has a "Strip Church" ministry in Las Vegas that reaches out to local sex industry workers. Strip Church reminds its target audience of strippers, prostitutes, porn stars, and card flickers (those people handing out advertisements along Las Vegas Boulevard for escort services) that someone cares for them, not through sermons or preaching, but through simple acts like offering cupcakes to dancers at strip clubs, food and water to card flickers, or a person to talk to. The church operates on the "belief that people deserve the love of Jesus Christ without the judgment of organized religion."[5] More mainstream churches, like a local Assemblies of God congregation, have ministries of women who greet dancers at strip clubs, give them gifts, and tell them that someone cares for them.

Pastor Jeff Howell of a nondenominational congregation in northwest Las Vegas, quoted Romans 5:20: "But where sin abounded, grace did much more abound." He added: "This is the city of grace. Have you ever seen a city with more grace? . . . The spiritual hunger is clear." Like Pastor Sears, Howell, who has lived most of his life in Indiana and Ohio, was hesitant about a call to serve in Las Vegas given what little he knew about the place. When a pastor friend approached him and asked if he wanted to join in starting a congregation in Las Vegas, Howell thought: "I had never wanted to visit. Nothing about Vegas appealed to me. Not the desert, not what I call the schlockyness of the place . . . the fake buildings, the fake plastic surgery on people's bodies." But, after he and his friend prayed about it for a month, they decided to pursue the idea and planned a four-day trip to the city. Howell recalled: "I was hooked. The spiritual needs were evident. The opportunity was evident [to build a church] in the fastest part of the fastest-growing city, at the very gates of hell itself." He stated that his goal was to come to Las Vegas and "seek and save the lost."

Jud Wilhite of Central Christian Church, another nondenominational congregation, had a similar uncertainty about the move to Las Vegas. But, like Howell, he has found that the many opportunities to spread the word and bring people to Christ make up for any concerns he had. In fact, as he wrote in his book *Stripped: Uncensored Grace on the Streets of Vegas,* he found that his own life experience as a recovering drug addict had prepared him for helping Las Vegans: "I know how it feels to be broken, to feel trapped to the point of hopelessness, to be numb to everything and ready to give up. Out of this experience comes my tremendous compassion for others who are struggling, and I believe this is part of why I ended up in Las Vegas." He invoked Paul's teachings on grace: "No matter how damaged we are by life, God's grace and love are only a turn away. That's why I don't call Vegas Sin City. I call it Grace City."[6]

Wilhite was only one of many Christian pastors who felt they had a "calling" or "mission" to heal the spiritually sick in Las Vegas. Elijah and Jonelle Randolph, husband-and-wife pastors at an African Methodist Episcopal church near downtown Las Vegas, saw the potential for service in the city after a vacation there in the early 1980s. They recalled the comments of people who criticized their move from Fort Wayne, Indiana, to Las Vegas to lead a local black church: "You're crazy. You're going to Sin City. You're going to Sodom and Gomorrah." Everyone is excited to go to Las Vegas on vacation, they added, but moving there was a different story. Elijah would respond to the disapproval with, "If it is as bad as you say it is, then maybe that's where we need to be as two ministers."

Pastor Glen Reardon, senior pastor at an Assemblies of God congregation, said he felt a special call in Las Vegas. He realized this while giving an invocation at a convention at the Mirage resort on the Strip. When some attendees brought up the topic of job security, Reardon offered to pray for them about their concerns. Then he told the group: "Speaking of job security. Can you think of any job with better security than mine? I'm called to help sinners, in Las Vegas. I'm always going to have a job." When he said that, he added, the group gave him a standing ovation. He continued: "What a great opportunity! . . . [In Vegas] you get a feeling you made a difference, that you did good." Another Christian pastor put such a calling this way: "We are on the front lines of sin in this town."[7]

JUST A REGULAR TOWN

Not all religious leaders I spoke with, however, felt theirs was a special call to reclaim sinners in Las Vegas. The majority of my interviewees, in fact, played down the Sin City image and made clear that their jobs here were the same as they would be anywhere else. For example, two Catholic priests explained that the city's unique qualities are what make it the same as other places. Father Antoine Pomeroi, who directs a campus ministry at UNLV, said that people often tell him: "Oh, your work is cut out for you [in Las Vegas]." He brushes aside such comments and explains that sinners are found everywhere and he sees no difference in his work here from his previous ministries at universities in Southern California, Tucson, Phoenix, and Eugene. Father George Toomey, who worked for Nevada Power in Las Vegas before becoming a priest, felt he had a call in Las Vegas, but not because it is more sinful than the rest of the world. Rather, he said, "Las Vegas is like anywhere else. Sin is sin and it's going to be anywhere."

Others felt similar about their work. Both Orthodox rabbi Levi Behar and Conservative rabbi Josef Rothschild explained that the Las Vegas location of their new positions didn't play into their choice to come. Rabbi Behar, in fact, noted specifically that he does the same things here that he would do anywhere else. Father Kent Sharp, a priest at an Eastern Orthodox church, said that his goal, regardless of location, is to work out his own salvation and then help other people through faith-based and service activities: "I don't feel a calling to go out and grab people out of the casinos. I'm here to be a light in the community." Reverend Tim Fuller, rector at an Episcopal congregation, agreed. When outsiders ask him, "Why go to Sin City?" he responds that Las Vegas has no more sinners per capita than Westfield, New York, where he used to pastor. Imam Azaan Hossein, a director at a local mosque, saw a unique opportunity in Las Vegas, but not necessarily to reclaim the sinners. When asked what he thought about his assignment, he responded: "I was excited in one sense. Because, if Muslims can express their religious identity in a place called Sin City, they can do it anywhere."

Even some pastors at conservative Protestant churches asserted the normality of Las Vegas religious culture. African Methodist Episcopal pastor Sergio Needham stated: "People always say 'Sin City.' You tell me what city is not Sin City. We just happen to have that label, that identity." He explained that his call to serve in Las Vegas, while special, is no different than any other and that he would teach the same thing here that he did in his previous church in Colorado. He quoted from the sixth chapter of Paul's epistle to the Romans: "For the wages of sin *is* death; but the gift of God *is* eternal life through Jesus Christ our Lord." Then he added: "My message is always of salvation. . . . I'm not making a paradigm shift."

Lutheran minister Ralph Merrill similarly recognized the uniqueness of Las Vegas but also viewed it as just another place where people need to hear the gospel: "I don't see Las Vegas as being a typically difficult place to do ministry. There is sin everywhere, but it is on a greater scale here. . . . The common denominator is that everyone has a need to be loved. Being in a place like Las Vegas the need is great. With gambling, drugs, and alcohol, people can hit rock bottom, and usually they will turn to the church." When asked if he felt a special call, Pastor Merrill summed up his feelings about the city: "Las Vegas, to me, wasn't a big deal. . . . It's another city where God's word and God's gospel need to be preached. And there's a lot of golf here."

TRIALS AND BLESSINGS

Despite claiming that Sin City is every city, nearly all cleric interviewees identified unique aspects of worship in Las Vegas. Some of these were challenges, others benefits, and most stem from the various city characteristics I have discussed in previous chapters.

One characteristic is diversity within individual congregations. Interviewees across the spectrum of faith traditions noted how their members come from many different backgrounds, an attribute of the overall city population. Orthodox priest Kent Sharp pointed to the ethnic diversity he has observed: "One of the things I love about this parish is that it is so eclectic." As evidence, he noted the many Russians who come to perform in Cirque du Soleil shows, a sizable refugee population from Eritrea, Sudan, and other African countries, additional American members, and more recent converts.

Reverend Tim Fuller's congregation is also ethnically diverse. He characterized it as mostly Caucasian American, with some African and African-American, Chinese, Korean, and Hispanic members, plus a Filipino membership so large that there is a special service for them each Sunday. Fuller explained that congregants come from a variety of religious backgrounds as well, a characteristic identified by several of his colleagues in other churches. Lutheran pastor Ian Sears, for example, noted that around half of his church's congregation is not Lutheran, but come from Catholic, Presbyterian, or Methodist backgrounds: "They are coming here for whatever reason. We are a large church in the area, but for some reason we attract a lot of non-Lutherans."

This diversity of Las Vegas religion is further illustrated in the mix of subgroups within some congregations. Raymond Drake, a director at the local Unitarian Universalists congregation, noted that his church community has created six different affinity groups in order for like-minded people to connect and help each other in individual spiritual searches. He compared this with his prior congregation in Washington DC, which had only two-to-four such groups. And, he added, such groups in Las Vegas "seem to coexist with few unpleasant interactions. In other areas, there is usually some friction."

All three rabbis I interviewed commented on a lack of entrenchment of Jewish culture in the relatively young and constantly changing city. Only recently have the local outlets existed to provide necessary food and services to practice their faith. Rabbi Rothschild explained how the city still has no Bureau of Jewish Education, which serves as a hub for a strong Jewish school system, or a Jewish old-age home. Rabbi Behar pointed to the city's growth. "This is a very young community," he said, noting that many Jews who came to Las Vegas to retire are not necessarily "looking for a spiritual life" or considering the availability of religion or synagogues. Rabbi David Banks agreed with Behar's assessment and pointed specifically to a lack of commitment to the next generation. He noted that around 40 percent of the city's Jewish population is retired, making youth education less of a concern than it might be in other more established cities. Banks added that since the existing population consists of a geographically diverse people, this further complicates the formation of an entrenched, forward-thinking community.

A diverse Jewish population can be an opportunity as well as a challenge. As the local community continues to grow, Rabbi Banks noted, it will eventually develop its own personality and social services. As one of the longest serving rabbis in the city, he was excited to have played a part in this growth so far and looks forward to the future: "How often can you say you built not one, but two temples?" Rabbi Rothschild also was optimistic about the potential: "It's exciting. You're kind of a pioneer in Las Vegas. Like I said, there is no infrastructure, so if you want to do something, you can make it happen." And, Rabbi Behar sees benefits in reaching out to the city's nonpracticing Jewish population: "There is a large Jewish community who is unaffiliated and need outreach." In fact, he added, Las Vegas has the largest such community within any city, "So, there's a lot of work to be done." It seems the Las Vegas "can-do" attitude extends to the realm of religion as well.

Several Catholic priests recognized the double-edged sword of growth. On the positive side, the church is constantly expanding and building new parishes, and their membership has the opportunity to build new traditions. According to Father Frank Green, around one-third of Las Vegas newcomers are Catholic, and the huge membership increases in the city often surprise colleagues in other dioceses. When one friend, a priest in the Juneau diocese, found out how many families attend Father Frank's parish, he remarked:

"You have more people in your parish than I have in my entire diocese." At the same time, the influx creates problems. Father Armando Sánchez noted: "We can't catch up with the growth. We need facilities. Sometimes our services are packed and we are probably breaking fire regulations." Father George Toomey mentioned another challenge of growth for the Catholic church. Because of such large congregations, he said, building "a one-on-one sense of community" becomes difficult, exemplifying another local trait.

Similarly, Mark Lewin, a bishop in the Church of Jesus Christ of Latter-day Saints, explained that because of the growing and transient population in the city, Las Vegans often look for a church near their house. Many times, he said, it doesn't have to be the same church they were attending before their move. So, on the upside, the number of Mormon converts in Las Vegas is generally higher than elsewhere in the United States. Then he added: "Because part of my ecclesiastical duties included shepherding the people," this growth becomes difficult and adds an additional responsibility and stewardship for church leaders and members.

The experiences of Lewin and others hint at an element related to the city's growth and diversity: many newcomers to Las Vegas do not mind *what* church they attend, so long as they can attach themselves to *a* church in their new home. Perhaps such a phenomenon is indicative of the "come to Las Vegas and start over" mentality of many transplants. Perhaps it indicates that church is one important source for community in a transient place where citizens often lack a sense of belonging. Pastor Elijah Randolph, for example, explained: "People come here to meet their neighbors." Such a hypothesis lends support to Frank Beckwith's explanation of the "more churches" myth; perhaps Las Vegas really does *need* more churches.

The city's transient nature also is reflected in many of the above comments and is a trait that adds to the challenges faced by local congregations. Father Kent Sharp observed: "You know, no one is really from here. If you've been here for twenty or thirty years, you're considered a native. So, there hasn't been the opportunity to establish generational connections. Back East you will find connections as deep as five or six generations. . . . It's hard to forge unity among people when their roots are so different." He added that it was only recently that his church had a member who grew up in the congregation join the parish council. Father Frank Green discussed the same challenges in his Henderson Catholic parish: "Because [our membership] is transitory, people don't take ownership of the church. They will say, 'This is not my church. Mine is back East.' [Or] they'll ask, 'Why build up the church.

I've already done this for my church.'" Such transience affects leadership too. Recall from chapter 3 how one Baptist minister struggled with keeping leaders in church positions because his congregation is such a "mobile community."

Pastor Ralph Merrill's comments about the city are representative: "That's the other thing about this town. It's a transient place. People are coming and going and coming and going and coming and going. We see that in the church too. . . . You know how a lot of seniors with mobile homes travel around the country and stop in different places? Las Vegas is one of those stops. And people stop into the church all the time."

Pastor Merrill gave another example of how religious organizations experience the city's mobility: "Because of the transient nature of the town, I have a lot of people calling in asking for money." Father Sharp admitted that a fellow priest in Phoenix has noticed a similar pattern, but said that in Las Vegas it will often be as many as five or six calls per week from people needing material assistance. Rabbi Rothschild said similarly that he receives calls at least once a week and sometimes every day from people asking for money, and pointed to the unique Las Vegas environment: "They all have a story, but in most of the cases I think they have a problem gambling. None of them admit it. They all have excuses."

Rabbi Behar said that calls to him for financial help were frequent as well, and added that most of the requests come not from members of his synagogue, but from outsiders who know that Chabad, his order of Orthodox Judaism, is known for assisting people in that manner. Imam Nassif, who also fields frequent requests from people needing help, has struggled with assessing the true intentions of requesters. Some people will call his masjid (mosque) and say, "Salaam Alaikum" (a common Islamic greeting, meaning peace be upon you) in order to give the impression they are part of the community before admitting, "I need a bus ticket out of town. I'm stuck." Nassif invites them to the masjid, where he can find out more about their situation (including whether or not they are truly Muslim), before helping them fill a gas tank or purchase a bus ticket.

A number of clerics discussed various other ways the Las Vegas 24/7 culture impacts their respective congregations. The large percentage of the local workforce who start their "weekends" on Tuesday or Thursday plays havoc with typical worship times. Father Sharp noted that whereas in other cities the parish hall would be more full for weekend services, that's not the case in Las Vegas: "The biggest challenge is work schedules. Because this is a seven-days-a-week, twenty-four-hour town, nobody has normal work schedules." Rabbi Banks agreed and pointed to the need for his synagogue to offer a service during the nighttime hours for graveyard-shift workers. Dr. Hank Taylor said that his Lutheran church is looking for an alternative worship schedule. He gave the example of one congregant who performs at a major production on the Strip, making it nearly impossible to attend a Sunday morning service. And, Dr. Hank added, he is not alone.

Pastor Merrill has instituted a midweek service with some success. He said: "One of the big things we deal with is people's schedules. You find that you have to adjust your worship times. When I came here [seven years ago] we had worship times on Sunday at 8:30 and 11:00 A.M. . . . We heard from some that 'We need a midweek service. I can't come on Sunday.'" As a result, his church began a Wednesday evening service, which has attracted between twenty-five and thirty attendees each week.

Father Antoine Pomeroi explained that many of the college students he shepherds have work schedules that don't permit them to come on Sundays. This is one of the biggest conflicts that he sees for Catholics in the Las Vegas environment. His solution is to encourage students with weekend work schedules to at least come to mass on Monday or Tuesday so that they don't completely lose their connection with their faith amid the demands of work and school. Away from the university, Father Armando Sánchez identified the same difficulty at his largely Latino Catholic parish.

Aside from preventing attendance at normal worship times, strange work schedules create challenges for the family, something churches hope to strengthen and build. Reverend Fuller, for example, explained the extra strain on families in his congregation that comes when parents work opposite shifts. Father Sharp made a similar statement and referenced the high number of latchkey kids in his parish. Bishop Mark Lewin summed up concerns about family in a city that never sleeps: "There's a change in the family dynamics because of the twenty-four-hour town."

THE WAGES OF SIN

The most obvious challenges for believers in Las Vegas revolve around the city's culture of vice. Even someone arguing for the normality of Las Vegas cannot avoid this unique element of the city. Consider Imam Hossein's view: "I think a lot is being said from all religions because this is called 'Sin City' and because of the gambling. Most congregants don't feel that. Temptation to do wrong is everywhere. Even in Jerusalem. In Jerusalem there are human rights abuses. There is prostitution in Jerusalem. There is gambling in Jerusalem. And that is the holiest of cities." When he referenced challenges specific to Muslims, however, he told a different story. Even though the same social and psychological challenges can be found everywhere, what is unique here is that "the Las Vegas economy is dependent on gambling, and that sometimes creates a dilemma for some Muslims. Gambling is forbidden in Islam." But Old Catholic Church bishop Dr. William Estes doesn't see a difference: "I don't think it is any different than any other city. . . . They have the same junk in Seattle and Portland as they do here. They have Indian casinos there, so there is gambling just like here. . . . They have the same topless bars in the big city as you do here. It's no different than anywhere else. It's just bigger, that's all."

The fact that sin is "just bigger" or more prevalent in Las Vegas was a key point made by twelve leaders I spoke with from the spectrum of faith traditions. Here are some other representative comments. Pastor Howell said bluntly: "Las Vegas has every temptation you can find elsewhere, but it is on steroids here." Bishop Lewin explained that, living "in the world," believers know they will be presented with temptations, "but here it's just more prevalent. You have easy access to things that cause a degenerative process to occur within the family." Pastor Sears compared Las Vegas to his former position at a midwestern church: "The only difference between Las Vegas and anywhere else is that it's not hidden here. It's just more open. In Cameron, Missouri, it's there! But, it's more hidden. It's almost like it is more accepted by [Las Vegans]."

When asked about challenges for his worshipers, Pastor Glen Reardon gave the following explanation: "As a man, you just get bombarded everywhere you go with temptation. They try to confine most of that to the Strip, but everywhere you go you have to keep your guard up. In other places, it's there but it's more subtle." He gave an example from his drive to his church the day we met. He looked out his side window and saw a photo of a nearly nude female on an adult entertainment newsrack advertisement. He recalled his thoughts at that moment: "There she is in all her glory, in one of those boxes."

Because of ever-present temptation, desensitization can become a problem for some believers. Imam Nassif said: "The greatest challenge [Muslims face in Las Vegas] is the overt, out-in-the-open lewdness I think most cities don't see on a regular basis." He added that, because of that openness, one can become numb to the effects of temptation: "That's the danger . . . when you don't see wrong as wrong." Rabbi Levi Behar explained the concern he has for his eight children, which could apply to religious people more generally: "It's not so much the slots in the stores. The kids get used to that. But it's the billboards on the back of taxis and the big ones too. . . . The kids get desensitized to seeing casinos. Unfortunately, they get desensitized to seeing scantily clad women on billboards."

The sexual imagery so prevalent on the Las Vegas landscape obviously is at odds with the teachings of most religions, causing some congregants to feel tension between their faith and the world around them. To understand how local religious groups see the struggle of moral faith in the context of the city's adult entertainment landscape, I asked interviewees how they might view a congregant's involvement in the sex industry. In nearly every case the response was generally, "love the sinner, hate the sin," an attitude summed up in a common Christian expression: "Church is a hospital for sinners, not a country club for saints." Such an attitude, of course, crosses into non-Christian teachings as well. Regarding believers who might have a problem with the forbidden act of gambling, Imam Quadir Nassif said: "You hate the sin and not the sinner. You don't kill the sick person."

Episcopal rector Tim Fuller said the church stands firmly against any illegal involvement in the adult industry, but added: "Being a liberal church, we stand welcome and would treat someone like that the same way as someone with a drinking problem or a gambling problem." Father Green similarly explained that, while the Catholic church looks down on involvement in pornography, it will accept the person involved in sin and will offer help: "We distinguish between the act and the person. We don't like the act, but . . . we love the person." Bishop Ted Burke expressed a similar welcoming attitude among Latter-day Saints. I like how Pastor Howell put it: "Come dirty with whatever dirt you have and God will clean you up."

Assemblies of God pastor Glen Reardon explained that, if there were strippers in his congregation, his response would be: "We're happy to have them. We pray for them. We have ministries for them." Even if they might be doing something the church interprets to be against what is taught in the scriptures, he wants them to feel welcome in the church. He added that hearing sermons against such activities might help effect change: "After a while, they are going to feel really uncomfortable. It's going to grate on them. We are trying to give a road map on the way to live, and they will realize after a while that 'this is probably not the way I need to live.'"

Some less-conservative churches welcome worshipers regardless of background and without requirement for change. Hank Taylor explained that he knows of strippers in his Lutheran congregation, but his position is one of grace. It doesn't matter if you're a stripper, he asserted. It doesn't matter if you're a bartender. Grace is when God looks at you and says, "I love you because I made you, because you're forgivable." "Ungrace" is when you have to do this or that to be accepted, he added. Dr. Hank explained that his is a philosophy derived from the New Testament scripture in Galatians 3:28: "There is neither Jew nor Greek, there is neither bond nor free, there is neither male nor female: for ye are all one in Christ Jesus." In Christ, nationality doesn't matter. Christ accepts all. "As a Galatians 3:28 church," he continued, "what we're lifting up is that grace is not only preached here, but practiced here. If a homeless man walked in, he'd be given as much of a welcome as Steve Wynn. In Christ it doesn't matter, so you are welcome."

Some interviewees explained that their approach to the tension is simply to teach and guide the flock away from sin and toward a more pure and faithful life. After a great deal of thought (and a very long pause) Old Catholic bishop William Estes answered my question about the tension between spiritual teachings and the Las Vegas environment as if he were giving a sermon: "There's always tension between right and wrong." He paused again: "But, when Christ died on the cross, he said, 'It is finished.'" Bishop Estes explained that we can accept His grace and His forgiveness, have a conversion, and be born into the Kingdom of God. He continued: "The scripture says we are set free from the power of sin and death. So, we don't have to do what the world does. We are not bound by sin anymore. If a person is living by the spirit of Christ, then there is no tension, unless he chooses to get involved with the world . . . unless he chooses to sin. Then that's where the tension starts. [The apostle] Paul said the things I want to do I can't, and the things I don't want to do I do. That is the natural tension between the flesh and the spirit."

Imam Nassif explained a similar lesson that he strives to teach his congregants from the perspective of Islam. He gave the example of several Muslim cab drivers attending his mosque. When they come to worship, Nassif asks them to park the taxis at a distance from the mosque because "they have naked women on them" and he wants to protect the other worshipers from such imagery. He mentioned the difficult situation such a job places believers in because they often take passengers to and from strip clubs. Even in such an environment, Imam Nassif explained, what is important is the heart and soul, and that one stay as far away from sin as possible. Every time one moves the boundary line through exposure or rationalization or desensitization, the danger becomes more imminent. To illustrate, he quoted the Muslim hadith, or statement from the prophet Mohammed: "If you have no shame, you may do what you please." Father George Toomey summed up the view of most ecclesiastical leaders I met. He explained that worshipers must "know the difference between what is labeled as okay in Las Vegas and what we should be thinking about as Christians."

When I asked clerical interviewees specifically about their views on gambling, I again found a relative uniformity of response. Aside from a handful of groups who expressly forbid gambling, the general feeling among religious leaders is that the practice is acceptable so long as it does not become a habit. Here is a sampling of representative viewpoints.

Jeff Howell explained how the Bible doesn't specifically proscribe gambling. He pointed to a passage from Paul's first epistle to the Corinthians where he teaches that "everything permissible, but not everything beneficial." Gaming is not condemned by scripture, Howell asserted. It comes closest to being a sin, he noted, in that gamblers ignore the value of work and seek to gain something that is not rightfully theirs: "It turns into a neutral, much like money." Money is not bad in itself, he explained, but you can do good or bad things with it. Drinking is another. He compared playing an eighty-dollar round of golf to going to a casino and spending the same amount on slot play and a buffet with a spouse. If gambling is done for entertainment value, and you have a budget for it, then it is not a sin. Rather, he cautions against the "abuse of and the addiction to" gambling.

Rabbis Behar and Rothschild took a similar position on the issue. Rabbi Behar explained: "[Gambling is] not really encouraged. If it's recreation and you're not gambling away your money, but doing it for entertainment, then that's different." Rabbi Rothschild said: "When you cross that line and spend money you don't have or that you can't afford to lose, then it becomes a problem." Both referred to greater principles they seek to teach: living within your means and not relying on charity, valuing money and work, not receiving something that is not earned, and involving oneself in a productive lifestyle.

Baptist pastor Dr. Melvin Roberts had a similar perspective about gamblers in his flock: "There is a difference in doing it playfully and doing it religiously. Doing anything to excess is sin." When asked how he might counsel someone who budgets gambling, he responded that he would focus not on sin, but on general principles similar to those taught by Behar and Rothschild: "I would tell them to reposition themselves for more satisfying entertainment, as I would an excessive spender. It's not a sin to be in debt, but it makes things difficult." For Dr. Roberts it comes down to teaching them to "sustain wealth and not throw it away."

From the Catholic point of view, Father Pomeroi compared gambling to drinking: "I always tell students that the church accepts the use, but condemns the abuse." But, he cautioned, "It is easy to be swept into the mentality of Las Vegas when you . . . live here." He added that he has learned of the detrimental effects of that addiction in confessions.

Elijah and Jonelle Randolph, like most of their colleagues, see gambling as inadvisable on a personal level, but teach moderation to their congregation. Elijah said he has had a number of people ask him recently about the church and gambling. "As a personal preference, it's probably better if you don't," he explained, "[but] I can't say it's a sin. . . . It's a good practice not to do it because it becomes habit forming." Jonelle added: "You can't say they are going to hell if they do it." Elijah summed up the views of a number of interviewees who denounce gambling only when it goes from budgeted entertainment to destructive habit: "When you get to the point when you use your bill money, that's when you cross the line. That's when it becomes a sin."

OPTIMISTIC FAITH

Amid the challenges to faithful worship in a growing, transient, and vice-ridden town, several religious leaders pointed out additional, specifically spiritual benefits that come with worshipping in the city. One such advantage relates to a spiritual need in the Las Vegas Valley, which interviewees felt fortunate to be able to meet. Dr. Gary Gable of one Methodist church in Henderson eloquently described his observations: "I see people who genuinely are seeking meaning in life." He said the reasons for their search often are the same ones people come to Las Vegas for in the first place: to start over, to run away from a problem, to make it big. "People come to Las Vegas who have huge needs," Dr. Gable continued. "It is the responsibility in church to hold up a mirror and say, 'here are the needs' and we show them some ways they can handle them." Peter Nickel described a similar situation at his nondenominational church: "I think people come here [to the church] for the same reasons people come to Vegas. They are hurting and have difficulty and want a new start. That's not everyone, of course. There are also those people who are searching for truth and answers. But, a lot of people come broken."

This blessing of service is, of course, evident in many of the comments I quoted earlier about feeling a "mission" to help the spiritually sick in Sin City. One such person, Pastor Jeff Howell, added a distinguishing second element of worship in the city that he considers a great joy. He compared his prior experience in Indiana and Ohio to Las Vegas: "In the Midwest it is more about [saving] face, about how you present yourself. . . . Here it's not that way. Here they come and dump their truck." He elaborated, telling a story from when he first arrived in Southern Nevada. A woman came to him and said: "I was molested as a child. I've been raped. I've had an abortion. But, what I want to talk to you about today is that my boyfriend is beating me." Pastor Howell noted that he has heard many such stories. He added: "In the Midwest it would have taken ten years of trust for someone to tell you one of those things. Here it takes five minutes."

Pastor Glen Reardon saw a similar pattern as he measured his experience in Las Vegas against his native Texas. Away from Southern Nevada, you might try to help someone make a connection or create a relationship with Christ in a public sermon, but few are likely to respond to an altar call, partly because so much of the population is religious. He noted that someone in Texas might say, "You're crazy if you think I'm going to raise my hand with Joe sitting right over there." In Las Vegas, however: "When I make an altar call, the hands go up. . . . When someone feels that conviction, people here will respond." The observations of Howell and Reardon, along with those of other pastors describing the hunger for spirituality in Las Vegas, underscore the candid, open character of the Las Vegas community I have discussed in previous chapters.

Another benefit identified by some clerics has to do with the spiritual strength that can be fostered in a city that presents overt opposition to religious teachings. A believer in Las Vegas has a distinct choice between sin and virtue. Recognizing that exposure to vice can also be a bad thing, especially if someone has a tendency toward one addiction or another, Catholic Father George Toomey said: "One thing about Vegas is that everything is out there and there's no pretension. . . . Nobody is pretending. It isn't hidden."

Both Latter-day Saint bishops I interviewed made similar comments. Bishop Burke acknowledged the negative side of exposure to gambling, which is strictly forbidden in that church, but added that people who recognize its presence and "fortify themselves against it and take a stand . . . tend to be stronger." He came back to the thought later in our discussion: "If you live here, it's kind of hard to have a run-of-the-mill lifestyle. You're either in the gambling, sex, and drugs side of things, or you're not. There's no middle ground."

Bishop Mark Lewin also spoke of such exposure, but related it to the

The Las Vegas Temple of the Church of Jesus Christ of Latter-day Saints. This edifice sits in a neighborhood at the foot of Frenchman Mountain overlooking the valley from its eastern edge. Photo by Dennis Rowley, March 2012.

young people in his congregation: "I also had stewardship and responsibility for the youth. I found it easier to teach them about good and evil" because of what Las Vegas is. He added that because of the blatant visibility in the valley of both the Strip and the Las Vegas Temple—one of many such edifices around the world that are considered to be the pinnacle of worship for Latter-day Saints—the distinction is readily apparent. "There is such a contrast that you can delineate more easily between good and evil." He acknowledged that the church's youth do not necessarily need to be in the temple all the time, nor do they need to stay away from nongambling casino amenities (bowling, movies, etc.) all the time, but the contrast between the two indicates a clear and present "symbol and a standard that helps them choose."

Imam Quadir Nassif's characterization of the black-and-white nature of sin and virtue in the city is particularly poignant. He spoke about the openness of sin in the community, calling it "a negative and a plus." I asked, why a plus? He answered: "A plus because the evil, or sin, is not creeping up on you. It's out in the open." So, he added, it is so easy to see that you can avoid it: "You don't have to just say, 'Beware of the boogeyman.' Here's the boogeyman right here."

IN THE WORLD BUT NOT OF THE WORLD

Given the powerful image of Las Vegas as Sin City, it is no wonder that misconceptions exist about the city's religious landscape. Women in Virginia will doubt the presence of churches in the city, Muslims outside of the city will wonder how Islam can thrive here, and outsiders will continue to ask local clerics if they have slots in their religious edifices.

Focusing on the religious community as it actually exists in Las Vegas clears up many of those misconceptions and simultaneously teaches about the city's sense of place. The answer to the broader question of how people of faith survive in this modern-day Babylon may be found in the principle that believers must live in the world but not of the world. One can remain a person of faith while living in a den of sin, but doing so requires a level of tolerance and negotiation between spiritual foundation and the surrounding context. Gambling is a good example of how such negotiation occurs.

Pastor Elijah Randolph explained: "The average church will frown on gaming." But in Las Vegas gambling is such a part of the city's economy and culture that it has become more acceptable, in religious terms, than in other locales. Whereas not living *of* the world in some other place might mean simply not partaking of the particular indulgence of gambling, completely avoiding this activity in Las Vegas may not be possible. Additional involvement in gambling—whether direct or indirect—seems to be a necessary evil in this city. Some additional comments lend support to that hypothesis.

On a personal level, many religious leaders do not consider gambling acceptable, and they would likely preach against its ills in congregations located outside Las Vegas. But some clerics I interviewed admitted that they cannot speak out against gambling in this city without potentially alienating many of their congregants who make a living in the industry, a loss that could threaten the viability of the congregation. Pastor Ian Sears, for instance, explained his adjustment: "Dealing with the gambling issue . . . I've realized that I can't hammer at it all the time." He said that he still sees a need to teach about its evils, but focuses instruction on recognizing the "difference between moderation and excess." Rabbi Josef Rothschild explained that he once gave a sermon about gambling that was really about the fundamental principle of living a meaningful life, doing good things with the resources you have, and not throwing them way. But, he recalled, it did not go over well with those in his synagogue who worked in the industry. Pastor Sergio Needham said similarly that when he preaches against gambling and the "wages of sin," it doesn't register: "What you're trying to combat is the same income that is coming to your church to help the church grow." And Dr. Gary Gable explained bluntly that, while Methodists "don't believe in gambling," in Las Vegas religious leaders put blinders on and choose only to deal with overt

problems like a gambling addiction: "In order to stay viable in the community, we choose not to deal with the issue."

Employment in the industry is another way in which Las Vegans show an
implicit acceptance of gambling as a necessary evil. Echoing the comments
of several other clerics I interviewed, Dr. Melvin Roberts described employees in the casino business: "They are looked upon as anyone in any industry
in any other town." The comments of one (nonclergy) interviewee who grew
up the son of a casino insider illustrates this attitude. Jeremy Mont is often
asked, "You're dad is a devout Catholic and a pit boss. How do you reconcile that?" To which he responds, "There's nothing to reconcile. That's just the
way it is and it is acceptable."

Jobs in the industry are so ubiquitous, in fact, that the two groups most
fervently against gambling have faithful members in supportive positions
within casino establishments. Latter-day Saints are encouraged not to take
jobs as dealers, but Bishop Ted Burke pointed to other faithful Mormons he
has known who worked in jobs from carpentry at one Strip resort to upper
management at another. Imam Quadir Nassif mentioned some congregants
at his mosque who work as cab drivers, a group dependent on the casinos
and other adult-oriented businesses. His colleague Imam Hossein said the
question of how to reconcile faith and work frequently arises in his conversations with other Muslims. The answer, he explained, is often left to the individual: "When they ask me, I tell them to get out of that kind of job as soon as
you can if you feel that it is something you should not do."

Imam Nassif also explained the "varying responses" in Islam to such que-
ries. "[Some people will say that] it is OK as long as you are not serving alcohol or directly involved in gaming itself [as a dealer]. Others will say to stay
completely away from the evil." He concluded with a telling remark about the
spiritual compromise that takes place in Las Vegas: "I wish I could say, 'don't
go into that environment at all,' but the reality is you have kids to feed." As
one religious instructor explained, sacrifice is giving up something you want
now for something you want more in the future. Some pious Las Vegans, it
seems, put aside their deep feelings against gambling for the long-term goals
of providing for family, furthering a career, or, in the case of some religious
leaders, retaining worshipers.

My survey of religious perspectives in Las Vegas reflects many of the city's
personality traits uncovered in other portions of this research: transience,
growth, opportunity to start over, and a can-do attitude, to name a few. Most

of all, however, the impact (both positive and negative) of the powerful Sin City image—created for and marketed largely to tourists—is another illustration of the bifurcated nature of the city. Specifically, the compromises made by local members of various faith traditions are another representation of the distinct but negotiable line between the insider's and outsider's sense of place. Notably, one (nonclergy) interviewee referred to the religious expression of "living in the world but not of the world" when she described what she would want an outsider to know about simply living in Las Vegas.

The city's religious experience also can teach a more generalizable lesson about place. People everywhere, in big cities or small, espouse aspects of religion or belief and, at the same time, live among people, objects, and philosophies that oppose these teachings. Believers are therefore forced to choose or compromise. After all, as several interviewees mentioned, every place can be a sin city. Dr. Melvin Roberts's comparison, I think, applies beyond the lights of Las Vegas: "Even though it is called Sin City, you must remember that, even in Rome in the days of old, there were saints in Caesar's palace. And, there are saints in Caesars Palace today."

Las Vegas Becoming

ntering the second decade of the twenty-first century, Las Vegas stands
at the brink. This is nothing new for the city that molds itself into whatever attracts the most visitors. Because of such a huge reliance on tourism, however, the city's economy has suffered disproportionately in the Great
Recession. The slumping economy only amplified huge questions locals
faced earlier regarding water availability, sense of community, rapid growth,
transience, and a movement toward a more economically diverse, less tour-
ism-dependent city. The result is a collective sense of introspection among
residents and observers about the future of Las Vegas. Will the city recover—
in visitor volume, casino take, construction employment—to a point when
growth rates maintain a strong positive trend, and people come again by the
thousands to fill all the now-empty homes? Will the foreclosure crisis, and
the screeching halt it brought to home construction, force locals to rethink
the sprawling urban growth trend of the past six decades? Will Las Vegas
become a city of its own, less tied to the Strip and tourism and more rooted
in arts, culture, and entertainment options for locals?

In returning to Las Vegas several times since completing interviews for
this project, I often pondered such questions. In my contemplation I have
added another query: What do the characterizations of the city I have
described in this book say about how it will react to current circumstances?
Predicting the future is as uncertain as a toss of the dice at the craps table,
but my observations from a recent winding drive through the valley provide
a glimpse into how unchanging elements in the local Las Vegas personality
may influence how the city confronts the precipice of its future.

A GLIMPSE OF REALITY

On a sunny day in January 2011, I drove north on Las Vegas Boulevard and
witnessed recent changes to this city of constant flux. CityCenter, the mammoth, multibillion-dollar MGM Resorts complex completed in 2009 at center Strip, is Las Vegas at its booming peak: extravagant, under construction,

flashy, and ready for the next stage of life as an upscale, hip, and global resort destination. I think of the project's ill-fated timing. In planning and early construction of the largest single resort endeavor in Strip history, everything in the city seemed economically perfect. Financing was easily secured, visitor volumes were steadily rising, and tourists were spending massive sums of money on expensive hotel rooms and condos, extravagant shows, and exorbitantly priced meals from celebrity chefs. By CityCenter's opening, however, all that had changed; the resort struggled with filling rooms, selling condos, and keeping gamblers on the casino floor.[1]

Passing Bellagio and Caesars Palace on the left, I am reminded what these two properties looked like when I was a teenager. Bellagio replaced the Dunes, and Caesars Palace is barely recognizable (aside from the legacy of a few Roman statue re-creations at the resort's gateway) with its numerous hotel tower additions that, like CityCenter, represent a more prosperous time. In that same reminiscent attitude I recall how Caesars' neighbor to the north, the Mirage, was the impetus for the resort and population explosion of the 1990s. Steve Wynn no longer runs that property, but his newest and highly successful ventures at Wynn Las Vegas and Encore another mile north remind me of his continued influence on the city's rebranding as a high-end, luxury tourism market. In fact, it is unlikely that MGM Resorts would have tried something as extravagant as CityCenter had Wynn Las Vegas been a failure.

Wynn's Encore, ironically, is where my vision of this newest act of the city's performance ends. Directly across the Boulevard, the former site of the Frontier is abnormally empty, a deserted parcel that was the most expensive tract ever sold in the city and that was supposed to accommodate another swanky high-brow resort.[2] Next to it, the stalled and silent construction site of Echelon Place, Boyd Gaming's attempted foray into the luxury Strip market. Gone are construction cranes that formerly were such a ubiquitous part of the Las Vegas skyline that Nevada historian Guy Rocha gave them the title of "state bird."[3] All that remains is a concrete shell awaiting a rebirth.

I continue northward, passing Circus Circus and the Riviera and another construction site, also depressingly silent but nearly complete. The Fontainebleau was to be yet another magnificent hotel-resort-condominium complex, but instead is frozen in time. Its most useful quality now: a place for the Clark County Fire Department to practice drills for high-rise emergency response.[4] My melancholy at seeing how the faltering local, national, and global economy has affected the city comes into sharp focus as I look across the street from the hulking blue shell of the resort and notice a sign for its defunct Career and Preview Center whose final letter has fallen from disrepair and neglect: "Guest Parkin."

At the Stratosphere, I turn west and weave my way through downtown. I first enter "18b," eighteen blocks the city has designated as an arts district. The Arts Factory at Charleston and Casino Center Boulevards anchors a smattering of struggling galleries that exist between tattoo parlors, lounges, coffee shops, and furniture stores. The First Friday art walk and street festival draws thousands to the area each month and engenders a unique sense of community, but the grand plan for this district has yet to materialize. Additional art establishments have materialized in other parts of the city's core. Some, like the Emergency Arts collective (housed in the former Fremont Medical Center building), have found a home northeast of the 18b arts district near Fremont East, another city-sponsored redevelopment district with a growing concentration of hipster bars, cafes, and clubs that is becoming a popular hangout for locals.

Leaving the Arts District, I take Bonneville Avenue as it dips west beneath railroad tracks and emerges in a once neglected desert parcel that is now a major hub of renewal downtown. On one corner of Bonneville and Grand Central Parkway the huge, sprawling World Market Center is striking. This mammoth complex of buildings hosts twice-yearly furniture markets and is

First Friday crowds cross the intersection of Charleston and Casino Center Boulevards as they move between street parties and exhibits during this popular monthly promotion by the Las Vegas Arts District. Organizers have done an excellent job working with city officials and gallery owners to attract more than ten thousand people each month from all corners of the valley. Photo by author, June 2007.

The Frank Gehry–designed Cleveland Clinic Lou Ruvo Center for Brain Health is a symbol of change in downtown Las Vegas. Photo by author, January 2011.

the envisioned hub of the home furnishings industry for the western states. The Las Vegas Premium Outlets and the Clark County Government Complex sit on opposite corners and together bring thousands of locals to the area each day. On the intersection's final corner is the Cleveland Clinic Lou Ruvo Center for Brain Health. This stunning Frank Gehry–designed structure opened in 2009 and has attracted attention beyond the city and not solely for its unique, eye-catching architecture. I park my car and explore the serene interior of the building, noticing people mingling as they attend some sort of professional meeting. Indeed, it has already become a local landmark and will play a global role in research and treatments for neurological disorders.

The Ruvo Center sits in the crook of a sixty-one-acre reclaimed Union Pacific Railroad brownfield, now reemerging as Symphony Park, a mixed-use urban community. The sights and sounds of construction, so eerily absent on the Strip, permeate the scene. Dozens of builders bustle at the site where the Smith Center for the Performing Arts rises from the desert floor. When opened, it will be host to many local performing arts events and the permanent home of the Las Vegas Philharmonic, Nevada Ballet Theater, and Broadway Las Vegas theater series. The city's Discovery Children's Museum also will occupy a space in the Smith Center complex, moving from its home in the same building as the Las Vegas Library on Las Vegas Boulevard, north of Fremont Street. The contrast between this site and the Strip is palpable; Symphony Park is moving and alive while the empty shells I saw on the Boulevard minutes earlier were lifeless and ghostly quiet.

I am awed by the transformation of this parcel of land in just a few short years. That change will only continue as more Cleveland Clinic facilities are added and other projects slated for the rest of Symphony Park come online, including a boutique hotel, an urban park, and high-rise residential buildings.

Visible from Symphony Park, just a block east of the railroad, I see another busy construction site that similarly breaks the silence felt on Las Vegas Boulevard. A new city hall for the city of Las Vegas is being erected at the corner of Main Street and Lewis Avenue. On this day, its form is similar to the shell of Echelon Place on the Strip, but its future more certain.

I leave downtown to explore the city's western suburbs on the final leg of my trip around the city. In years past, I would have expected to see construction vehicles moving in and out of emerging neighborhoods, but that scene, instead, is replaced with an abnormally large number of for sale signs. A few minutes later I notice the steel skeleton of the unfinished Shops at Summerlin Centre mall near the Las Vegas Beltway and Charleston Boulevard, a reminder that the same forces stalling projects on the Strip affect locals too.

Toward the end of my tour, I drop in on some longtime friends. Our discussion inevitably turns to the recession's impact on home prices. They complain that even though they have been paying the mortgage on their home for over twenty years, they owe more than its appraised value since they had taken so much out of the home's rapidly rising equity during the boom years. Although theirs is a common story, just as typical is the family that purchased a home as prices were skyrocketing in hopes of huge equity returns and were left with an overblown mortgage when the housing bubble burst. My friends plan to stick it out, but so many other locals have left their underwater mortgages, moving away from Las Vegas or to a rental property elsewhere in the city. Indeed, the Great Recession has cut deeply into the suburbs.

A WINDOW INTO THE FUTURE

This tour of the city is a snapshot of reality and simultaneously a lens into the city's potential future. More specifically, three perspectives on recent economic struggles reflect the city's character, developed over decades, that will continue to influence Las Vegas for years to come.

Initially evident in my tour of the city is a reemphasis of a local trait and common refrain in this book: that life in Las Vegas is not all that different from life in other parts of the country, but is amplified in certain aspects. Across the country cities and towns have felt the weight of the Great Recession, and every American knows someone who has lost their job, left their home, or, at the very least, sold a house for much less than it was worth mere months earlier. Locals in other cities have watched as construction projects stalled because of financial difficulties—from the Waterview Tower condominiums on Chicago's lakeshore to the planned renovation of an empty, neglected downtown Houston hotel.[5] But, true to form, Las Vegas has been on the bleeding edge of that struggle.

The city's foreclosure and unemployment trends bear this out. One out of every nine local homes went into foreclosure in 2010, making Las Vegas the number one foreclosure city in the country that year. It held the same ranking in 2009, a move up the list from fourth in 2008 and fifth in 2007. In short, Las Vegas no longer fits the title of Fastest Growing City in the Nation, but instead holds the more dubious one of Foreclosure Capital. Home values in the city average less than 40 percent of what they were at the top of the boom in 2006, dropping to levels not seen since the mid-1990s. And unemployment numbers rose to a record 14.9 percent at the end of 2010, more than five points above the national rate. Furthermore, a consistent rise in Las Vegas casino gaming take will follow an overall uptick in economic conditions nationwide since gaming numbers are based largely on discretionary income, in scant supply during a recession.[6]

Secondly, what I saw on my tour clearly demonstrates the city's bifurcation into local and tourist parts, another core trait of the city's personality. While stalled hotel and resort construction projects depend on increases in tourism revenue to bring them back to life, lively cultural institutions under construction will depend largely on locals for their survival. I have noted a number of times in previous chapters one tenet of Las Vegas: as the Strip goes, so goes the rest of Las Vegas. Historically, that is a true statement. Such contrasting scenes in my tour indicate, however, that at least some of what goes on in the local's side of the city may point toward a future that is largely separate from that of the tourist side.

Amid the severe impact of the Great Recession on the city, community leaders and local observers have feverishly suggested broadening the city's economic base beyond gaming and cultural offerings beyond headliners or Cirque du Soleil shows, all the while pushing toward sustainability in urban growth. Of course, such discussions have been in the background for years, even decades. But, the need for change often becomes more acute for people faced with extreme trial. Talk of diversification similarly becomes more widespread each time Las Vegas faces an economic downturn. Thus, Las Vegas may be on the verge of a new, alternative trajectory epitomized in the projects I saw under way in downtown Las Vegas.

Before 18b or the Smith Center began to emerge, in fact, downtown Las Vegas was not really a downtown at all, at least not in the mindset of many local transplants from the East or Midwest. It was certainly the municipal and government services core, but in terms of culture and entertainment, downtown typically has been considered a lesser tourist attraction, after the Strip. Aside from a few locals who loyally patronize the downtown casino and restaurant establishments, most would not come to the area for entertainment as they might in other cities. In fact, until recently many locals perceived the environs surrounding Fremont Street as rundown, dirty, and crime-ridden.

That began to change toward the latter end of the boom years, and continued in the wake of the Great Recession. Symbolizing such change was the razing of the neglected Queen of Hearts hotel-casino, which had become a haven for drugs and prostitution. In its stead will be the new Las Vegas City Hall, a hopeful impetus for a mixed-use development in the neighborhood surrounding the government complex.[7] Nearby, the Arts District—while by no means a cohesive establishment in location or intention—has moved forward, in fits and starts, through hard times. That the arts community has continued to open galleries and maintains several geographic nodes throughout the downtown core testifies to the district's budding, persistent, and progressive nature. The Fremont East entertainment district adds a hipster hangout, a key component to bringing people downtown. Just south of Fremont East, the uniquely "Vegas" Neon Museum and other attractions in the city's Cultural Corridor continue to draw both locals and tourists downtown. And when city employees move into the new city hall, the old building complex will be occupied by a large and growing company in the Las Vegas Valley. Zappos, the Internet shoe and clothing warehouse, has purchased it and plans to move its corporate headquarters from Henderson to the former city hall space and hopes some of its thousands of employees will take advantage of the growing live-work culture in downtown.[8]

The Lou Ruvo Center for Brain Health and its collaboration with the Cleveland Clinic, a world-renowned leader in brain and heart care, signals the emergence of a medical research and services component in the local economy. The center could add a much different element to the global image of Las Vegas. Its founder, Larry Ruvo, son of the center's namesake, foresees the city as "a Mecca for health care," and one of the slogans for Keep Memory Alive, the Ruvo Center's fund-raising arm, boasts: "What happens in Vegas will benefit the rest of the world."[9]

The Smith Center for the Performing Arts is the flagship in the current iteration of cultural development downtown. When I met with Don Snyder, chairman of the organization's board of directors, and Myron Martin, president and CEO, more than five years before its opening, they referred to the Smith Center as "the living room for Las Vegas." Martin referenced the can-do attitude so prevalent in the city and asserted that the Smith Center will be a tool to build community and business in the downtown corridor and "the impetus for arts and culture" in Las Vegas. "A rising tide lifts all boats," he continued. "We will be that tide, lifting the boats up with arts and education." Snyder added that it will be "the most important series of buildings built in our lifetime. . . . [It will impact] more broadly and deeply than any other institution in the community." He boldly compared its impact to that of Boulder Dam, which saved Las Vegas from the worst of the Great Depression. The art-deco structure, designed by architect David M. Schwarz, is even reminiscent of the look of the dam, making Snyder's comparison all the more appropriate. The Smith Center, he boasted, is the "catalyst" for so many development and redevelopment efforts in the city and, he implied, will provide the "momentum" to help the city "get out of this economic ditch."[10]

Struggles with residential foreclosure, lowered home values, and halted construction away from the city center, however, continue parallel to stalled construction and diminishing returns on the Strip. Thus, a third perspective on the Las Vegas character and future is based on how the local's side of the city remains connected to its tourist counterpart in fundamental ways.

Las Vegas is, along with the rest of the nation, adjusting to a "new normal" in the economy, one where anticipated growth occurs at a much lower rate than during the boom times. CityCenter is a (literal) shining symbol of the overreaching, extreme expectations about what was possible previously, and the silenced construction on the Strip in 2011 is mute testimony to the unrealistic nature of such prospects. A well-respected local analyst, Jeremy Aguero, put it this way: "We can't grow like we grew before. . . . We're going to have to change our expectations a little bit. As the economy starts to shift, this becomes the new reality."[11] Still, in suggesting that Las Vegas faces a "new normal" for growth, it is significant that *growth* is still expected, it just takes on a different face. On the Strip, managers at properties such as the Tropicana and the Stratosphere, for example, have been focusing on renovations rather than new projects, and operators at Treasure Island have targeted a middle market rather than following the highbrow trajectory of the Strip before the bottom fell out of the economy.[12]

The actions of residents throughout Las Vegas will undoubtedly reflect trends on the Strip. Just as the economy created ripple effects in tourism, the housing crisis extended to nearly every part of life in the suburbs, from stalled projects such as the Shops at Summerlin Centre to a shuttered Las Vegas Art Museum to decreased funding for public education. As a result, locals may moderate their aspirations for bigger and more, and may exchange the desire for a dream home (previously based, of course, on anticipated huge equity returns) for one they can afford. Dropping real estate prices, in fact, suggest that Las Vegas is on a return to a "more normal," less inflated economic state, similar to what existed just a decade earlier, when service workers seeking the American Dream on a middle-class wage and baby-boomers looking for an affordable place to retire could find fulfillment in Southern Nevada.

Similarly, as several Strip resort operators have done, Las Vegans may need to step back to retool their focus moving forward. What might that retooling look like? It may come in the form of examining (with the rest of Nevada) the legacy of Western-style individualist, libertarian, low-tax political culture so pervasive in the state's history. Such a paradigm shift in political leadership may help community leaders to identify new ways to bolster deficient education and health-care systems to assist in diversifying the culture and economy and thereby boost the overall quality of life in the new normal. Another part of this shift must include the way Las Vegas leaders and residents look at the sustainability of water. Even though the city has been praised for its efforts in conservation, growth without more supply will, of course, limit realistic future prospects. Furthermore, if the current supply from the Colorado basin (or the potential supply from central Nevada aquifers accessed via a proposed pipeline) runs dry, this desert metropolis bent on growth may join the ranks of other boomtowns in the state and dwindle to ghost-town status.

A revised outlook for Las Vegas also may engender a new focus on building the sense of community suppressed for so long by the city's growth and transience. Indeed, hard times and the struggle to keep up a neighborhood pocked with vacant homes may draw remaining residents together. And

Wasted runoff in a typical Las Vegas neighborhood near Flamingo and Sandhill Roads. Water problems will forever plague Las Vegas as population growth outstretches the supply available through an allotment from the Colorado River. Yet, simple conservation efforts such as water-efficient plants and irrigation methods could lessen the waste tremendously. Photo by Dennis Rowley, April 2012.

dropping home values keep many residents (such as my friends who owe so much on their mortgage) from selling their homes, thus locking them down in their neighborhood. This could actually foster a less mobile population and may help to strengthen community bonds through increasing longevity. Steve Friess, a longtime local writer, explained his optimism: "The people who hang on to their houses and rebuild from the wreckage that's around us, those folks will have a greater sense of ownership over their community. They will feel like Las Vegans in a way that you don't when there is no adversity."[13]

In my observations of the city in flux, however, I do not see a change in overall character. Rather, I see a Las Vegas on the path of maturity. Las Vegas is, after all, the youngest big city in the country, and as a "youthful" place, it is prone to make some mistakes, to meet a few bumps in the road. Perhaps it grew too big too fast. Perhaps locals saw the easy-financing/housing-construction bubble as more opportunity than risk and got burned. During the boom times, locals often commented how the city was on the verge of becoming something more than just a tourist town. It may be that the difficult circumstances Las Vegans faced during the recession will magnify this desire and make it more possible. I see the situation akin to that moment for many young adults when they face a crisis that forces them to grow up, take responsibility, and grapple with the struggles of adulthood. As the city enters its second hundred years of life and leaders and residents ask themselves what they, collectively, want to be when they grow up, perhaps the wake of the Great Recession will be a similar moment. Just as the characteristics of a young child can lead her to overcome challenges in her adult years, so will the Las Vegas character of the past influence its reaction to a challenging future.

LAS VEGAS PERSONALITY

The timing of this book—from the period of initial research to the subsequent years of writing—has straddled a pivotal point in the lifespan of Las Vegas. As things go in this city, it seems this project is just another example of a walk down the two-way street with lady luck. In a way, I was fortunate to have conducted my interviews when I did; Las Vegas had not yet fallen into the economic slump. If my interviews had taken place later, I am afraid they would have lacked a perspective and depth, and the optimism that characterizes Las Vegas. At the same time, because they occurred near the high point of the Vegas boom, my interviews also fail to capture the deep turmoil and difficult circumstances locals have felt since the economy tanked and foreclosures skyrocketed. Yet, focusing on the character so identifiable in the good times, together with how that character may influence the city through the bad, gives a deeper, more holistic portrayal of the local's sense of identity and place.

Las Vegas has been a city of opportunity, change, growth, and negotiation between the outsider's image and local life throughout its history. Thus, the transformation the city has faced during the recession may be new in substance, but not circumstance. Las Vegas has always been a growing, can-do town, one that is moldable, changeable, and flexible. But, the core personality I have sketched in this book has not changed. Throughout its history, Las Vegas has capitalized on the opportunities that arose: from Boulder Dam in the late 1920s and early 1930s, to the postwar tourism boom, to the rapid economic expansion of the late 1980s and early 1990s and the bubble of the early 2000s. Because of such a dependence on external forces for its lifeblood, the city's culture and sense of identity reflect such influences, the most predominant today, of course, being tourism and the "Vegas image." The two forces at work as the city enters its second hundred years of life—financial crisis and a need to mature for its two million residents—are simply another impetus for the next wave of change. As it has before, the city and its residents will adjust and build a place that will, in turn, lead to something better down the road. The Vegas spirit will endure and see the city through challenges and choices to a brighter future as a grown-up city. As Dr. Melvin Roberts, a Baptist minister in West Las Vegas, observed: "What Las Vegas offers is a melting pot of positives and negatives, possibilities and deterrents, [and] it depends on how you stir the pot and how you foster a community of nurturing a future beyond Las Vegas Boulevard." Las Vegas locals will find a way to stir the pot to keep their city glimmering in the desolation of the Mojave Desert.

NOTES

Abbreviations

LVRJ   *Las Vegas Review-Journal*
LVS   *Las Vegas Sun*

Introduction

1. Landor Associates and Penn, Schoen, and Berland Associates, "2006 Newsmaker Brands," December 27, 2006; Benjamin Spillman, "Google, Las Vegas Top Brands in '06 and '07, Survey Finds," *LVRJ*, January 11, 2007.

2. Las Vegas Convention and Visitors Authority (LVCVA), "Historical Las Vegas Visitors Statistics, 1970–2009," http://www.lvcva.com/.

3. John Western, *A Passage to England: Barbadian Londoners Speak of Home* (Minneapolis: University of Minnesota Press, 1992); Yi-Fu Tuan, *Space and Place: The Perspective of Experience* (Minneapolis: University of Minnesota Press, 1977); Kenneth E. Foote, *Shadowed Ground: America's Landscapes of Violence and Tragedy*, rev. ed. (Austin: University of Texas Press, 2003).

4. Robert R. Riley, "The Visible, the Visual, and the Vicarious: Questions About Vision, Landscape, and Experience," in *Understanding Ordinary Landscapes*, ed. Paul Groth and Todd W. Bressi, 200–209 (New Haven, CT: Yale University Press, 1997). Quotation from page 207.

5. D. C. D. Pocock, "Place and the Novelist," *Transactions of the Institute of British Geographers* 6, no. 3 (1981): 337–47. Quotation from page 337.

6. Kent C. Ryden, *Mapping the Invisible Landscape: Folklore, Writing, and the Sense of Place* (Iowa City: University of Iowa Press, 1993), 40–41, 43.

7. Tuan, *Space and Place*, 72–73.

8. Ibid., 18.

9. Ibid., 198.

10. Applied Analysis, *2007 Las Vegas Perspective* (Las Vegas: Metropolitan Research Association, 2007), 7.

11. Jane Jacobs, *The Death and Life of Great American Cities* (New York: Random House, 1961), preface.

12. John K. Wright, "Terrae Incognitae: The Place of Imagination in Geography," *Annals of the Association of American Geographers* 37, no. 1 (1947): 1–15. Quotation from page 1.

ONE : One Hundred Years of Opportunity, Luck, and Rapid Change

1. Quoted in Stanley W. Paher, *Las Vegas, as It Began—as It Grew* (Las Vegas: Nevada Publications, 1971), 16. Drivers along Interstate 15, which today follows the same general path, would do well to remember this warning before venturing off the freeway without a good supply of water, food, and fuel, especially during the summer months. I blew a tire along this stretch of the *jornada de muerto* a couple of years back. Though my donut replacement tire proved to have air, the anxiety I felt as I searched for it gave me a small taste of what travel along this road was like in days past.

2. Applied Analysis, *2007 Las Vegas Perspective*, 30.

3. Ralph J. Roske, *Las Vegas: A Desert Paradise* (Tulsa, OK: Continental Heritage Press, 1986), 31.

4. Hal G. Curtis, oral interview, Las Vegas: The Las Vegas I Remember Series, Nevada State Museum and Historical Society and KNPR Studios, 1988.

5. Elbert Edwards, qtd. in Andrew J. Dunbar and Dennis McBride, *Building Hoover Dam: An Oral History of the Great Depression* (New York: Twayne, 1993), 16.

6. Leon Rockwell, manuscript of oral history, Leon Rockwell Papers, MS13, Department of Special Collections, Lied Library, University of Nevada, Las Vegas, 125–26.

7. "Give Thanks at Boulder Site," *Las Vegas Age*, December 22, 1928.

8. "From Where I Sit," *LVRJ*, January 1, 1932.

9. Valleywide census numbers include Las Vegas and North Las Vegas for 1940 and add in the total for Henderson for 1950. Unfortunately, numbers for unincorporated Clark County within the Las Vegas Valley are not available, so these totals reflect a smaller population than actually existed in the valley at that time.

TWO : A Place in the Desert

1. Bill Fiero, *Geology of the Great Basin* (Reno: University of Nevada Press, 1986), 127.

2. Ibid., 134–35.

3. Frenchman Mountain is often mistakenly called Sunrise Mountain by locals, but the latter is farther north and smaller than the former. The confusion is understandable. The more noticeable of the two peaks is what Las Vegans see the sun rise above each morning.

4. Sherman Frederick, "The 'Dry Heat' of Las Vegas," *LVRJ*, June 24, 2007.

5. Eugene P. Moehring, *Resort City in the Sunbelt: Las Vegas 1930–2000*, 2nd ed. (Reno: University of Nevada Press, 2000), 230–32.

6. Royal Embassy of Saudi Arabia Web page, http:// www.saudiembassy.net/.

7. Valaer Murray, "America's Most Visited Cities," *Forbes.com*, April 28, 2010; Valaer Murray, "America's Top Tourist Attractions," *Forbes.com*, May 20, 2010.

8. Applied Analysis, *2007 Las Vegas Perspective*, 77; Las Vegas Convention and Visitors Authority (LVCVA), *Economic Impact Series: The Relative Dependence on Tourism of Major U.S. Economies* (Las Vegas: Applied Analysis, 2010).

9. Airport statistics gathered from http://www.mccarran.com/.

10. Antonio Planas, "Clark County School District: Hispanic Students Outnumber Whites," *LVRJ*, November 18, 2006.

11. No relation to the author.

12. Quoted in Wilbur S. Shepperson, ed., *East of Eden, West of Zion: Essays on Nevada* (Reno: University of Nevada Press, 1989), viii.

THREE : Watch 'Em Come, See 'Em Go

1. John Katsilometes, "Famed Designer Willis Never Tires of, 'What's Your Sign?'" *Las Vegas Weekly*, June 9, 2008.

2. John M. Beville, "How Las Vegas Pioneered a Rotary Club," *Nevada Historical Society Quarterly* 10, no. 3 (1967): 29–34. Quotation from page 31.

3. A. D. Hopkins, "Ed Von Tobel (1873–1967)," in *The First 100: Portraits of the Men and Women Who Shaped Las Vegas*, ed. A. D. Hopkins and K. J. Evans (Las Vegas: Huntington Press, 1999), 40–41.

4. Jerry Fink, "A Hard Road for a Dream," *LVS*, July 21, 2006.

5. Ibid.

6. Ibid.

7. R. Marsh Starks, "Life's a Gamble," *LVS*, October 15, 2006.

8. Lynette Curtis, "Report: Nevada Has Highest Percentage of Homeless in U.S.," *LVRJ*, January 11, 2007.

9. *The Word on the Street: Homeless Men in Las Vegas* (Reno: University of Nevada Press, 2005), 67.

10. Ibid., 67.

11. Matthew O'Brien, *Beneath the Neon: Life and Death in the Tunnels of Las Vegas* (Las Vegas: Huntington Press, 2007), 20–26.

12. Ibid., 25.

13. Borchard, *Word on the Street*, 122, 148.

14. O'Brien, *Beneath the Neon*, 25.

15. Ibid., 74.

16. Marshall Allen, "A Poor Place to be a Kid," *LVS*, July 23, 2006.

17. A sheriff's card, or work card, is required for many of the service jobs in the city, particularly those in the casino business. Seen as a remnant from former mob days, the work card is issued after a check of the criminal background of potential employees and certifies their eligibility to work under city, county, and state regulations.

18. Mike Trask, "Seniors' Dreams Can Fade Away," *LVS*, December 31, 2006.

19. Brendan Buhler, "Consider All Things, Stations Ask," *LVS*, October 20, 2006.

20. Robert Futrell, Cristie Batson, Barbara G. Brents, Andrew Dassopoulos, Crissy Nicholas, Mark J. Salvaggio, and Candace Griffith, *Las Vegas Metropolitan Area Social Survey: 2010 Highlights* (Las Vegas: Department of Sociology, University of Nevada, Las Vegas, 2010), 28.

21. Emily Richmond, "This New Teacher Plans to Buck the Odds, and Stay," *LVS*, March 10, 2008.

22. Wallace Stegner, *Where the Bluebird Sings to the Lemonade Springs: Living and Writing in the West* (1992; New York: Penguin Books, 1993), 138, 201.

23. See, for example, John McPhee, *The Pine Barrens* (New York: Farrar, Straus & Giroux, 1967); Western, *A Passage to England*; Ryden, *Mapping the Invisible Landscape*; J. B. Jackson, *A Sense of Place, a Sense of Time* (New Haven, CT: Yale University Press, 1994), 159–162; Keith H. Basso, "Wisdom Sits in Places: Notes on a Western Apache

Landscape," in *Senses of Place,* ed. Steven Feld and Keith H. Basso (Santa Fe: School of American Research Press, 1996), 53–90; William Least Heat-Moon, *Prairyerth (A Deep Map)* (New York: Mariner Books, 1999); Cary de Wit, "Women's Sense of Place on the American High Plains," *Great Plains Quarterly* 21 (2001): 29–44.

FOUR : Getting Along with Growth

1. I found that at least two versions exist for this cover of *Life:* the one with the caption I quote here and another that says instead, "LAS VEGAS—BIGGER AND LIVE-LIER." "Gambling Town Pushes Its Luck," *Life,* June 20, 1955, 20–27.

2. "Too-Fast Cities: Five Spots Where the Risks Outweigh the Upside," *Fast Company,* issue 117, July 2007; Jennifer Robison, "Report Says LV Growing Too Fast," *LVRJ,* July 12, 2007.

3. Center for Business and Economic Research, *Southern Nevada Factbook,* Spring 2006 (University of Nevada, Las Vegas, 2006), 1, 12. Thanks to Bob Potts at UNLV's CBER for pointing out this pattern to me.

4. Henry Brean, "Clark County Population: 2 Million," *LVRJ,* December 5, 2007.

5. US Department of Education, National Center for Education Statistics, "Common Core of Data," http://nces.ed.gov/ccd/; Mark Gottdiener, Claudia C. Collins, and David R. Dickens, *Las Vegas: The Social Production of an All-American City* (Malden, MA: Blackwell, 1999), 100; Anthony Ramirez, "Sign of the Times: Smaller Class of New Teachers," *LVS,* August 19, 2010; Clark County School District, "Fast Facts 2010–2011" and "Fast Facts 2007–2008," http://ccsd.net/.

6. Emily Richmond, "What's in a Name? Often It's Contention," *LVS,* September 16, 2008.

7. Antonio Planas, "Teacher Shortage: School District in 'Crisis'," *LVRJ,* July 28, 2005; Emily Richmond, "Puttin Out the Welcome Mat," *LVS,* July 18, 2006; Emily Richmond, "Teacher Void Starting to Fill," *LVS,* July 16, 2007; Emily Richmond, "District Still in No Rush to Fill Teacher Vacancies," *LVS,* July 14, 2008.

8. Robert W. Burchell, Anthony Downs, Barbara McCann, and Sahan Mukherji, *Sprawl Costs: Economic Impacts of Unchecked Development* (Washington DC: Island Press, 2005), 174–77.

9. John G. Edwards, "Growing Debate: Will It Be Up or Out?" *LVRJ,* December 25, 2005.

10. Samantha Young, "Spreading Out: Suburban Growth Has Price," *LVRJ,* November 14, 2005; Hubble Smith, "Home Growth Going Rural," *LVRJ,* September 14, 2006; quotation from Edwards, "Growing Debate" (see note 9).

11. Gottdiener, Collins, and Dickens, *Las Vegas: The Social Production,* 135.

12. Henry Brean, "Las Vegas Wash: Chunks of Past Build Future," *LVRJ,* October 3, 2005; Kristen Peterson, "With Rally, Fans Will Show Love of Midcentury Modern, Frazier Hall," *LVS,* November 28, 2007; Kristen Peterson, "Neon Treasures, Not Trash," *LVS,* December 22, 2006; Joe Schoenmann, "Fifth Street School Soon to Brim with Students of the Arts," *LVS,* May 29, 2008.

13. Alan Choate, "'Pop' Squires' House Faces Wrecking Ball," *LVRJ,* December 1, 2010; Lynnette Curtis, "Kiel Ranch Still the Worse for Wear," *LVRJ,* July 20, 2007; Mike Trask,

"Restoration Outlook Is Dreary," *LVS*, January 23, 2009; Joe Schoenmann, "Owner Says He's Tried to Find Use for Huntridge," *LVS*, March 7, 2008; Cy Ryan, "State to Help Owner Find New Use for Old Huntridge Theatre," *LVS*, March 26, 2008; Alan Choate, "Historic House Looking for a Home," *LVRJ*, November 8, 2010.

14. Charlotte Hsu, "Take a Long Look at Frazier Hall," *LVS*, July 26, 2008; Peterson, "With Rally" (see note 12).

15. Brean, "Las Vegas Wash" (see note 12).

16. Peterson, "Neon Treasures" (see note 12); Kristen Peterson, "Just a Little Bit Longer, Baby," *LVS*, February 19; Kristen Peterson, "Neon Boneyard Park Still Not Quite Open to Public," *LVS*, May 13, 2011. Quotation from Kristen Peterson, "Signs of Old Times," *LVS*, August 27, 2006.

17. David McGrath Schwartz, "Las Vegas Ready to Make History," *LVRJ*, August 2, 2006; Alan Choate, "LV Museum Gets $800,000 in New Funds," *LVRJ*, November 27, 2007; Alan Choate, "Mob Museum Unveils Barber Chair from 1957 Murder," *LVRJ*, March 9, 2011.

18. David McGrath Schwartz, "Historic School Getting New Life as Work of Art," *LVRJ*, January 6, 2006; Kristen Peterson, "Lessons in History," *LVS*, May 26, 2007; Schoenmann, "Fifth Street School" (see note 12); City of Las Vegas, "Historic Fifth Street School," http://www.lasvegasnevada.gov/.

19. Charles F. Carpenter, "Letter: Construction Takes Away More Natural Beauty," *LVS*, June 8, 2007.

20. John L. Smith, "Gilcrease Family's Decision to Sell Understandable but Still Heartbreaking," *LVRJ*, July 3, 2005; Adrienne Packer, "Housing Plan Backed on 40 Acres at Orchard," *LVRJ*, September 8, 2005.

21. "Cost of Living Index for Selected U.S. Cities," http://www.infoplease.com/ipa/A0883960.html.

22. Sam Skolnik, "Hall of Justice Can't Catch Up to Growth," *LVS*, October 16, 2006; Geoff Schumacher, "For Library District, A Penny Saved Is a Penny Earned," *LVRJ*, April 2, 2006.

23. Cristina Silva, "Nevada's Boom Ends in Record Number of Empty Homes," *LVS*, March 15, 2011.

24. Silva, "Nevada's Boom" (see note 23); Jackie Valley, "Foreclosures, Bad Economy Create Fertile Ground for Marijuana 'Grow Houses,'" *LVS*, November 12, 2010.

25. Buck Wargo, "Foreclosed Homes a Bane to Neighbors in Las Vegas Valley Communities," *LVS*, December 10, 2010.

26. Kathy Williams, letter to the editor, *LVS*, April 4, 2008.

27. Ronald W. Smith, ed., *The Las Vegas Metropolitan Area Project: Studies of Selected Quality of Life Issues* (Las Vegas: University of Nevada, Las Vegas, 1998); Terri Lynn Hicks, "Quality of Life Amidst the Lights of Las Vegas" (master's thesis, University of Nevada, Las Vegas, 1999); Hal Rothman, "Too Big for Our Britches?" *LVS*, October 9, 2005; R. Keith Schwer, Sandra Phillips Johnson, Rennae Daneshvary, Richard W. Hoyt, and Saadullah Bashir, *2007 Southern Nevada Community Assessment* (Las Vegas: Center for Business and Economic Research, University of Nevada, Las Vegas/United Way of Southern Nevada, 2007); Nevada Chapter of the American Institute of

Architects, *Blueprint for Nevada* (Las Vegas: AIA Nevada, 2008); Futrell et al., *Las Vegas Metropolitan Area Social Survey: 2010 Highlights*.

28. Lisa Mascaro, "Nevada in Line for a Seat," *LVS*, January 3, 2007; Steve Tetreault, "Nevada Gains Seat in U.S. House of Representatives," *LVRJ*, December 21, 2010.

29. Editorial, *Las Vegas Age*, January 8, 1929; editorial, *Las Vegas Age*, March 14, 1929.

30. Applied Analysis, *2007 Las Vegas Perspective*, 6, 26, 27, 53, 65, 74, and 77.

31. Mike Davis, "House of Cards," *Sierra* 80, November/December 1995: 36–42.

32. Joe Schoenmann and Ed Koch, "Rogers Dares Utter Dreaded 'Income Tax'," *LVS*, April 26, 2007; Bruce Ruger, letter to the editor, *LVRJ*, October 31, 2006; J. J. Nowakawski, letter to the editor, *LVS*, March 4, 1997; Bill Smith, letter to the editor, *LVRJ*, July 17, 2007; "Mayor Says City Must Control Unmanaged Growth," *LVS*, February 9, 1998; Evan Blythin, letter to the editor, *LVRJ*, November 29, 2007; Pat Cole, letter to the editor, *LVS*, July 19, 2008; Mike Trask, "Looking in On: Suburbs," *LVS*, November 6, 2007. Quotation from Trask article.

33. Hubble Smith, "Development: Give My Regards to Main Street," *LVRJ*, December 3, 2006.

34. Richard Velotta, "In Business Q and A: Oscar Goodman, Las Vegas Mayor," *In Business Las Vegas*, May 4–10, 2007.

35. Tim Richardson, "Nevada Leads Nation in Rate of Foreclosures," *LVS*, July 25, 2008; Howard Stutz, "Low and Behold," *LVRJ*, May 10, 2008; Chris Morris and Alex Richards, "Growth-related Jobs Nose-dive," *LVS*, August 14, 2008; Sam Skolnik, "Signs Are Dim, but Is Vegas' Future?" *LVS*, August 3, 2008; Henry Brean, "Golden Gate's Shrimp Cocktail Special a Sign of the Times," *LVRJ*, May 4, 2008.

36. Howard Stutz, "The Boulevard Formula," *Las Vegas Business Press*, February 5, 2008; Liz Benston, "How Many New Rooms? Let's Just Say, a Lot," *LVS*, February 16, 2008.

37. Henry Brean, "Another Million Neighbors?" *LVRJ*, December 9, 2007.

38. "Boyd Gaming Suspends Construction Work on Echelon Project," *LVRJ*, August 1, 2008; Liz Benston, "Flailing Economy Cited in Boyd's Decision to Postpone Further Construction of Echelon Project," *LVS*, August 1, 2008; Jennifer Robinson, "Southern Nevada Economy: Analysts' Projections Sour," *LVRJ*, August 2, 2008; Liz Benston, "Echelon Isn't Alone in Not Making the Cut," *LVS*, August 11, 2008; Tamara Audi, "Las Vegas Plaza Plans Pushed Back to 2009," *Wall Street Journal*, August 15, 2008; Richard N. Velotta, "Executive: No Plans in Place for Stalled Fontainebleau Project on Las Vegas Strip," *LVS*, February 9, 2011; Steve Green, "Sahara's Closure on May 16 Will Mark 'The End of an Era'," *LVS*, March 11, 2011.

FIVE : Stuck in the Fast Lane

1. Omar Sofradzija, "Route 160: Deadly Road Gets Attention," *LVRJ*, February 28, 2006; Omar Sofradzija, "Crosses Serve as Stark Reminder of Route 160 Carnage," *LVRJ*, March 5, 2006.

2. Omar Sofradzija, "Speed Limit Set for Route 160," *LVRJ*, March 3, 2006; Omar Sofradzija, "Roadside Memorials: Taking Up the Cross," *LVRJ*, March 11, 2006; quotation from Sofradzija, "Crosses Serve as Stark Reminder" (see note 1).

3. Sofradzija, "Deadly Road Gets Attention" (see note 1); Sofradzija, "Speed Limit Set" (see note 2); Jack Bulavsky, "A Safer Highway," *View News*, December 29, 2009; Jason Gardner, letter to editor, *LVRJ*, March 6, 2006; K. Hansen, letter to editor, *LVRJ*, March 7, 2006; quotation from Gene Kraft, letter to editor, *LVRJ*, March 9, 2006.

4. Omar Sofradzija, "Driving Out of Sync in 2007," *LVRJ*, January 7, 2007.

5. John Przybys, "New Campaign Uses Everyday People Who Have Received Traffic Violations to Promote Safe Driving," *LVRJ*, November 15, 2005.

6. Brendan Buhler, "Even with a Map, You Can Still Get Lost in Las Vegas," *LVS*, August 23, 2006; Clark County GIS Management Office, "Downloadable GIS Data Sets," http://gisgate.co.clark.nv.us/.

7. Regional Transportation Commission of Southern Nevada, *Regional Transportation Plan: 2009–2030* (Las Vegas, NV: RTC, 2008), ES3, 17.

8. David Schrank, Tim Lomax, and Shawn Turner, *Urban Mobility Report 2010* (College Station: Texas Transportation Institute, 2010), 34–35, 45; American Association of State Highway and Transportation Officials, *Transportation: Invest in Our Future— America's Freight Challenge* (Washington DC: AASHTO, 2007), 19–20; Jennifer Robinson, "Businesses Troubled by Traffic," *LVRJ*, February 7, 2007.

9. Andrew Kiraly, "Why Can't Las Vegas Drive?" *Las Vegas CityLife*, February 8, 2007.

10. Tom Wagner, letter to the editor, *LVS*, January 22, 2007.

11. Edward Murphy, letter to the editor, *LVRJ*, March 20, 2006.

12. Omar Sofradzija, "Gripes and Groans Galore," *LVRJ*, February 25, 2007.

13. Quentin R. Aukeman, letter to the editor, *LVRJ*, September 9, 2006.

14. Tom Gorman, "On Viewing Road Bullies as Social Misfits," *LVS*, January 15, 2006.

15. Dr. Michael Pravica, letter to the editor, *LVS*, June 27, 2008; editorial, *LVS*, June 21, 2008.

16. John Devine, letter to the editor, *LVS*, July 3, 2008.

17. Bob Coffman, letter to the editor, *LVS*, January 29, 2008.

18. Penny Rice, letter to the editor, *LVS*, January 31, 2008.

19. Carol Nguyen, letter to the editor, *LVS*, January 31, 2008.

20. Phil McKay, letter to the editor, *LVS*, January 31, 2008.

21. R. G. Aldrich, letter to the editor, *LVRJ*, December 15, 2006.

22. Sofradzija, "Driving out of Sync" (see note 4).

23. Andrew Kiraly's article "Why Can't Las Vegas Drive?" in *Las Vegas CityLife* (see note 9), inspired this part of my discussion.

24. Omar Sofradzija, "Comedy of Errors Results in Tragedy of Route 160," *LVRJ*, June 11, 2006.

25. Sofradzija, "Crosses Serve as Stark Reminder" (see note 1); Sofradzija, "Deadly Road Gets Attention" (see note 1); Sofradzija, "Speed Limit Set" (see note 2); Omar Sofradzija, "Route 160 Speed Limit Lowered," *LVRJ*, March 7, 2006.

26. Editorial, "Carnage on Blue Diamond Road," *LVRJ*, March 8, 2006; Sofradzija, "Comedy of Errors" (see note 24). Quotation from Bulavsky, "A Safer Highway" (see note 3).

27. Editorial, "Driving into Gridlock," *LVS*, December 10, 2006; Ed Vogel, "Transpor-

tation Official Says Road Projects Need More Cash," *LVRJ,* February 20, 2008; Omar Sofradzija, "Highway Funding: Road Bill Pleases Gibbons," *LVRJ,* June 8, 2007.

28. Jacob Snow, "Jacob Snow Talks About the Steps the RTC Is Taking to Keep Las Vegas Traffic Moving," *LVS,* August 30, 2007; Francis McCabe, "Officials Discuss Revenue Shortfall for Road Work," *LVRJ,* February 20, 2008.

29. Sofradzija, "Deadly Road Gets Attention" (see note 1).

30. Sofradzija, "Route 160 Speed Limit Lowered" (see note 25).

31. David Kihara, "Mean Streets," *LVS,* March 19, 2006.

32. Kiraly, "Why Can't Las Vegas Drive?" (see note 9).

33. Kihara, "Mean Streets" (see note 31).

34. Chuck Styles, letter to the editor, *LVS,* February 1, 2008.

35. Editorial, *Las Vegas Evening Review,* May 6, 1929; editorial, *Las Vegas Evening Review,* May 18, 1929.

36. Nguyen, letter (see note 19).

37. Kiraly, "Why Can't Las Vegas Drive?" (see note 9).

38. Ibid.

39. Las Vegas Convention and Visitors Authority (LVCVA), "2007 Las Vegas Year-to-Date Executive Summary" and "Las Vegas Visitor Profile, Calendar Year 2007, Annual Report" (prepared by GLS Research), http://www.lvcva.com/, 26; Adrienne Packer, "Strip's Growth Causing I-15 Woes," *LVRJ,* February 22, 2006.

40. Hal Rothman, *Neon Metropolis: How Las Vegas Started the Twenty-First Century* (New York: Routledge, 2002), 249–51; Las Vegas Convention and Visitors Authority (LVCVA), "2007 Las Vegas Year-to-Date Executive Summary," and "Las Vegas Visitor Profile, Calendar Year 2007, Southern California and International Visitors Version," 1, 30.

41. Kihara, "Mean Streets" (see note 31).

42. Omar Sofradzija, "Newbie Finds Wild West on Valley Roads," *LVRJ,* November 19, 2006.

43. Jane Ann Morrison, "Never Mind the Cheating Spouses—We Need Traffic Cameras to Protect the Drivers," *LVRJ,* March 24, 2007.

44. Sofradzija, "Gripes and Groans Galore" (see note 12).

45. E-mail communication with Toni Pond, crime analyst, Traffic Bureau, Metropolitan Police Department, July 23, 2007, Las Vegas, NV; Michelle Ernst, Marisa Lang, and Stephen Davis, *Dangerous by Design* (Washington DC: Transportation for America, 2011), 11.

46. Gorman, "On Viewing Road Bullies" (see note 14).

47. Kiraly, "Why Can't Las Vegas Drive?" (see note 9).

48. Kihara, "Mean Streets" (see note 31); Kiraly, "Why Can't Las Vegas Drive?" (see note 9).

49. Schrank, Lomax, and Turner, *2010 Urban Mobility Report,* 34–35 (see note 8); Omar Sofradzija, "Commuting Times: Study: Pay Now for Roads," *LVRJ,* August 31, 2006; editorial, "Driving into Gridlock" (see note 27).

50. Brookings Institution, "Vehicle Miles Traveled (VMT)," 2008, http://www.brookings.edu/.

51. Allstate Insurance Co., "The 2010 'Allstate America's Best Drivers Report'" Complete National List," http://www.allstatenewsroom.com/. Allstate data does not include any cities in Massachusetts.

52. National Highway Traffic Safety Administration, "State Motor Vehicle Crash Statistics," http://www.nrd.nhtsa.dot.gov/; Sofradzija, "Newbie Finds Wild West" (see note 42).

53. Editorial, "Carnage on Blue Diamond Road" (see note 26).

54. Kihara, "Mean Streets" (see note 31).

55. Phillip Miynek, letter to the editor, *LVRJ*, August 23, 2007.

56. Gorman, "On Viewing Road Bullies" (see note 14).

SIX : Locals in a Tourist City

1. Brian Frehner, "'Squeezing the Juice out of Las Vegas': Reflections on Growing up in Smalltown, USA," in *The Grit Beneath the Glitter: Tales from the Real Las Vegas,* ed. Hal K. Rothman and Mike Davis (Berkeley and Los Angeles: University of California Press, 2002), 189.

2. Ibid., 188–90.

3. Shannon McMackin, "I Didn't Know Anybody Lived There," in *The Grit Beneath the Glitter,* 202–3.

4. Kathleen Hennessey, "Resort Corridor: Pool Crashing Elevated to Art," *LVRJ,* September 5, 2006.

5. McMackin, "I Didn't Know Anybody Lived There," 203, 206.

6. Frehner, "Squeezing the Juice," 192.

7. Bruce Spotleson, "Locals Remain a Unique Species," *In Business Las Vegas,* March 19, 2010.

8. Mark Hansel, "Child Care Fills 24/7 Needs," *LVS,* June 28, 2006; "20 Answers: How Is Business Unique in Las Vegas?" *Vegas Inc.,* April 18, 2011.

9. Jack Sheehan, "Why Las Vegans Can Be a Tad Defensive About How Others See Their City," *LVS,* April 13, 2008.

10. April Corbin, "Home Means Las Vegas: Natives Offer Up Their Unique Perspectives on the City's Evolution," *Las Vegas Weekly,* January 13, 2011.

11. Emily Richmond, "Stranded Students," *LVS,* December 28, 2007.

12. Tom Gorman, "On an Innovative Program to Convince Youths the Value of Their Education," *LVS,* November 18, 2005.

13. Annie E. Casey Foundation, "Kids Count Data Center," Teens Ages 16 to 19 Not in School and Not High School Graduates (Percent)—2009, http://datacenter .kidscount.org/; Lynnette Curtis, "Kids Count Data Book: Study Paints Grim Picture," *LVRJ,* June 27, 2006; Emily Richmond, "For Third Straight Year, More County Kids Stay in School," *LVS,* March 6, 2007; Emily Richmond, "Dropping Out to Go to Work," *LVS,* May 15, 2008. Quotations from Hubble Smith, "Marketplace: Useful Pursuits," *LVRJ,* February 12, 2007.

14. UNLV Institutional Analysis and Planning, "Six-Year Graduation Rates of New Freshmen, Fall 2003 Cohort," http://ir.unlv.edu/; NCHEMS (National Center for Higher Education Management Systems) Information Center, "Progress and Completion:

Six-Year Graduation Rates of Bachelor's Students," http://www.higheredinfo.org/. Western states included in the average include Nevada, California, Oregon, Washington, Idaho, Utah, Arizona, New Mexico, Colorado, Wyoming, and Montana.

15. Charlotte Hsu, "Valley College Students Create Their Own Work-Study," *LVS*, March 10, 2008; Ed Koch and Emily Richmond, "Big Issues Are All About People: Education," *LVS*, November 5, 2006.

16. Marshall Allen, "It's a Myth, No Park It," *LVS*, August 27, 2006; Corey Levitan, "Crash Course: A Lot of Trouble," *LVRJ*, May 21, 2007. Quotation from Allen, "It's a Myth."

17. Liz Benston, "The Card Dealer," *LVS*, November 19, 2007; Charlotte Hsu, "Student's Plan Illustrates Quirk in Local Job Market," *LVS*, March 5, 2008; Hsu, "Valley College Students" (see note 15).

18. Ibid.

19. John Przybys, "Variety of Influences: What Makes Las Vegas Las Vegas," *LVRJ*, May 29, 2008.

20. Marshall Allen, "Friendly Neighbors Often Waved Off," *LVS*, July 10, 2006.

21. Przybys, "Variety of Influences" (see note 19).

22. "New City Park to Be Named to Honor Siegfried and Roy," *LVRJ*, July 20, 2005.

23. Tiffany Brown, "More Than a Mere Marathon," *LVS*, December 4, 2007; David McGrath Schwartz, Sonya Padgett, and Todd Dewey, "New Las Vegas Marathon: Vegas Kitsch Flavors Race," *LVRJ*, December 11, 2006.

24. Jack Sheehan, "Jack Sheehan Says Tourism Slogan Accurately Describes a Visitor's Feelings About Las Vegas," *LVS*, February 26, 2006.

25. Sonya Padgett, "Unleashing the Inner Exotic Dancer: Once Around the Pole," *LVRJ*, July 13, 2006; Jerry Fink, "Looking in On: Entertainment," *LVS*, July 6, 2007.

26. Tom Gorman, "On Questioning the Mayor's Flamboyant Behavior and Its Impact on the City's Image," *LVS*, January 1, 2006.

27. Corey Levitan, "Locals Don't Bat an Eye: Only in Las Vegas," *LVRJ*, November 2, 2006; John L. Smith, "Prohibition-era Mayors Make Oscar Look Like Small Potatoes," *LVRJ*, May 17, 2005; David McGrath Schwartz, "Mayor of Mixology," *LVRJ*, February 28, 2007; Mark Hansel, "Over the Years, The Mayor Has Been a Man of Many Faces, But Just One Name: Oscar," *LVS*, April 10, 2007; Alan Choate, "Goodman Considers His Political Future," *LVRJ*, January 17, 2009.

28. Kelli Schlueter, "R-Jeneration: R-Voice," *LVRJ*, September 26, 2006.

29. Hansel, "Over the Years" (see note 27).

30. John L. Smith, "Sage Advice from a Strip Mall Par for the Course in Our City of Surprises," *LVRJ*, June 5, 2005.

31. Emily Kumler, "Valley Pearls," *LVRJ*, July 21, 2005.

32. "Rafael Villanueva: The Vegas Guy," *BLVDS Las Vegas*, May 2007, 13.

SEVEN : Life in a Town of Glitter and Gold

1. Las Vegas Convention and Visitors Authority (LVCVA), "2010 Clark County Resident Study" (prepared by GLS Research), http://www.lvcva.com/, 13, 16, 29, 36, 37;

LVCVA, "2008 Clark County Resident Study" (prepared by GLS Research), http://www.lvcva.com/, 41–42.

2. Adam Goldman, "Paycheck Cashing at Casinos Questioned," *LVS*, November 8, 2004.

3. R. Marsh Starks, "Life's a Gamble," *LVS*, October 15, 2006.

4. Goldman, "Paycheck Cashing" (see note 2).

5. National Gambling Impact Study Commission (NGISC), "Final Report" (Washington DC: National Gambling Impact Study Commission, 1999), 4–9.

6. Rachel A. Volberg, "Gambling and Problem Gambling in Nevada: Report to the Nevada Department of Human Resources" (North Hampton, MA: Gemini Research, Ltd., 2002), 40.

7. NGISC, "Final Report," 4–6.

8. Volberg, "Gambling and Problem Gambling in Nevada," iii, 29, 32; NGISC, "Final Report," 4–5.

9. William N. Thompson and R. Keith Schwer, "Beyond the Limits of Recreation: Social Costs of Gambling in Southern Nevada," *Journal of Public Budgeting, Accounting and Financial Management* 17, no. 1 (2005): 85.

10. Abigail Goldman, "You Bet There's a Problem," *LVS*, July 27, 2006; Arnold M. Knightly, "Problem Gambling Bill Faces No 'Real Opposition,'" *LVRJ*, May 22, 2007; interview with Sam Skolnik on KNPR, "The High Stakes of Gambling," *KNPR's State of Nevada* podcast, July 6, 2011 (Las Vegas: Nevada Public Radio).

11. Rachel A. Volberg, "Gambling and Problem Gambling Among Adolescents in Nevada: Report to the Nevada Department of Human Resources" (North Hampton, MA: Gemini Research, 2002), 16–17, 30.

12. Kathryn Hausbeck, "Who Puts the 'Sin' in 'Sin City' Stories? Girls of Grit and Glitter in the City of Women," in *The Grit Beneath the Glitter: Tales from the Real Las Vegas*, 340–41.

13. Shari L. Peterson, "Suggestive Billboards: Should They be Tamed?" *LVS*, April 3, 2004.

14. Ibid.; and Gary Peck and Allen Lichtenstein, "Suggestive Billboards: Should They Be Tamed?" *LVS*, April 3, 2004.

15. Carri Geer Thevenot and Henry Brean, "Free Speech," *LVRJ*, July 13, 2007; Carri Geer Thevenot, "Brothel Billboard Cruises Paradise Road," *LVRJ*, September 1, 2007; Bob Herbert, "City as Predator," *New York Times*, September 4, 2007; editorial, *LVS*, September 5, 2007.

16. Alan Choate, "Council Agenda Includes 'Only in Vegas' Items," *LVRJ*, October 14, 2007; Joe Schoenmann, "Looking in On: City Hall," *LVS*, November 25, 2007.

17. Joe Schoenmann, "'Stripper-Mobile' with Live Dancers Raises Safety, Decency Concerns," *LVS*, November 11, 2009.

18. Jon Ralston, "Jon Ralston Wonders Whether Rampant Gambling, Prostitution Should Be Part of the Image That Las Vegans Want to Create," *LVS*, September 9, 2007.

19. Joe Schoenmann, "Las Vegas Sells Sex, Sometimes Gets Coy," *LVS*, November 21, 2007; Peterson, "Suggestive Billboards" (see note 13); Schoenmann, "Stripper-Mobile" (see note 17).

20. "City Closes Clinic Tied to Unsafe Procedures," *LVRJ,* March 2, 2008

21. Ibid., comment section.

22. Lisa McKensie, letter to the editor, *LVRJ,* March 5, 2008; Carol Vick, letter to the editor, *LVRJ,* March 4, 2008; Mary Ashcraft, letter to the editor, *LVRJ,* March 5, 2008.

23. John Przybys, "Get Out of the Fast Lane: Artistic Diversions," *LVRJ,* April 2, 2006.

24. Marie Sanchez, "Growing Up in Las Vegas," in *The Real Las Vegas: Life Beyond the Strip,* ed. David Littlejohn (New York: Oxford University Press, 1999), 84.

25. Donald W. Meinig, "Symbolic Landscapes: Some Idealizations of American Communities," in *The Interpretation of Ordinary Landscapes,* ed. Donald W. Meinig (New York: Oxford University Press, 1979), 164–65, 174–76.

EIGHT : Religion in Sin City

1. Jud Wilhite, with Bill Taaffe, *Stripped: Uncensored Grace on the Streets of Vegas* (Colorado Springs, CO: Multnomah Books, 2006), 169.

2. Joan Whitely, "The Place to Pray," *LVRJ,* May 11, 1997; Association of Statisticians of American Religious Bodies home page, http://www.asarb.org/.

3. Wilhite, *Stripped,* 68.

4. Emmily Bristol, "Mission from God," *Las Vegas CityLife,* March 8, 2007; Christopher Lawrence, "Life on the Couch: 'Hookers: Saved on the Strip' Focuses on Hope, Redemption," *LVRJ,* December 5, 2007.

5. Sonya Padgett, "Sex Industry Ministry: XXXChurch.com Strives to Demonstrate Jesus' Love in Las Vegas," *LVRJ,* May 30, 2010; Mike Trask, "Minister to Base His War on Porn at Local Church," *LVS,* October 26, 2008.

6. Wilhite, *Stripped,* 16–18, 69, 124–25. Quotations from 125 and 69, respectively.

7. Lori Leibovich, "Houses of the Holy," in *The Real Las Vegas: Life Beyond the Strip,* 171.

Conclusion

1. Buck Wargo, "CityCenter Condo Closing Slow Down in Economy," *LVS,* May 28, 2010 ; Howard Stutz, "CityCenter's Value Plummets," *LVRJ,* August 4, 2010; Liz Benston, "Too Many Rooms to Fill: CityCenter's Opening Felt, Even as Visitor Volume Grows," *LVS,* August 9, 2010.

2. Hubble Smith, "Recession Leaves Raw Land in Dust," *LVRJ,* March 7, 2010.

3. Laura Myers, "With Boom Days Gone, Nevada's Future Offers Some Tough Choices," *LVRJ,* January 2, 2011.

4. Jennifer Robinson, "Firefighters Conduct Drills at Fontainebleau," *LVRJ,* January 28, 2011.

5. Eliot Brown, "New Life for Stalled Project," *Wall Street Journal,* July 27, 2011; Cara Buckley, "Deal on Stalled Condo Project Is First Under a City Program," *New York Times,* March 22, 2011; Purva Patel, "Bankruptcy Stalls Plans for Deserted Hotel," *Houston Chronicle,* August 3, 2011.

6. Buck Wargo, "Las Vegas Leads Nation in Foreclosures for 2009," *LVS,* January 27, 2010; Buck Wargo, "Las Vegas Records Nation's Highest Rate of Foreclosures in 2010," *LVS,* January 27, 2011; Jennifer Robinson, "Las Vegas Unemployment Rate Climbs to

Record 14.9 Percent," *LVRJ*, January 21, 2011; Hubble Smith, "Las Vegas Home Sales Up, Prices Continue to Decline," *LVRJ*, April 27, 2011; Derek Kravitz, "Home Prices Rise in Many U.S. Cities; Las Vegas an Exception," *LVRJ*, June 28, 2011; Hubble Smith, "Analyst: Gaming Cannot Save LV," *LVRJ*, October 6, 2009; Chris Sieroty, "Pain for Locals Is Pain for Gaming," *LVRJ*, July 20, 2011.

7. Dave Toplikar, "Queen of Hearts Demolition Marks New Era for Downtown," *LVS*, February 2, 2010.

8. Alan Choate, "Boosters See Zappos Move as Latest Step in Bringing People Back to Downtown," *LVRJ*, December 5, 2010; Joe Schoenmann, "Zappos CEO Envisions a New Community Downtown," *LVS*, March 17, 2011; Joe Schoenmann, "Zappos Brainstorming How to Make Downtown More Livable," *LVS*, May 20, 2011.

9. Larry Ruvo, "Vegas Will Draw Visitors, Residents Who Want High-Quality Health Care," *LVS*, December 27, 2009; Keep Memory Alive, main page, http://www.keepmemoryalive.org/.

10. Steve Green, "Smith Center Board Chairman: The Greatest Project Since Hoover Dam," *Vegas Inc.*, May 2, 2011; editorial, "Heart of the Arts: Smith Center the City's 'Living Room,'" *LVRJ*, March 11, 2012.

11. Hubble Smith, "Analysts Address LV's Lagging Economy," *LVRJ*, June 8, 2010.

12. Geoff Schumacher, "The Renovation Era," *LVRJ*, December 3, 2010; Amanda Finnegan, "Stratosphere Undergoes $20 Million Renovation," *LVS*, January 12, 2011; Liz Benston, "Tropicana's $180 Million Renovation Turning Heads," *Vegas Inc.*, May 27, 2011; Amanda Finnegan, "Phil Ruffin Shuns High End as He Modifies Treasure Island," *Vegas Inc.*, May 31, 2011.

13. KNPR, "The Once and Future Las Vegas," *KNPR's State of Nevada* podcast, June 24, 2011 (Las Vegas: Nevada Public Radio).

BIBLIOGRAPHY

INTERVIEWS BY THE AUTHOR

Abreu, Annie. June 1, 2007. Las Vegas, NV.

Allison, Trish. June 4, 2007. Summerlin, NV.

Andre. July 12, 2007. Las Vegas, NV.

Aponte, Edwin. March 19, 2007. Las Vegas, NV.

Arnett, Nola. June 5, 2007. Las Vegas, NV.

Ballard, Chuck. March 12, 2007. Las Vegas, NV.

Banks, David. July 24, 2007. Henderson, NV.

Bannister, Ernest. June 7, 2007. Las Vegas, NV.

Barnes, Drew. February 17, 2007. Las Vegas, NV.

Behar, Levi. July 13, 2007. Las Vegas, NV.

Boyne, Gary. June 1, 2007. Las Vegas, NV.

Bridger, Mike. March 23, 2007. Las Vegas, NV.

Brookings, Craig. January 20, 2006, and May 3, 2007. North Las Vegas, NV.

Bryn. March 5, 2007. Las Vegas, NV.

Burke, Ted. July 21, 2007. Las Vegas, NV.

Busch, Russell. March 29, 2007. Las Vegas, NV.

Cale. June 25, 2007. Las Vegas, NV.

Carlin. February 8, 2007. Las Vegas, NV.

Carlton, Misty. May 25, 2007. Las Vegas, NV.

Charles. January 28, 2006. Las Vegas, NV.

Cutter, Aaron. July 28, 2007. Las Vegas, NV.

DeAngelo, Ted. July 15, 2005. Las Vegas, NV.

Del Toro, Jimmy. June 27, 2007. Las Vegas, NV.

Dodson, Matthew. June 1, 2007. Las Vegas, NV.

Donald, Jeff. January 18, 2007. Las Vegas, NV.

Drake, Raymond. July 2, 2007. Las Vegas, NV.

Estes, William. June 20, 2007. Las Vegas, NV.

Fishman, Louise. March 1, 2007. Las Vegas, NV.

Foster, Beth. January 23, 2007. Las Vegas, NV.

Foster, Shirley. January 23, 2007. Las Vegas, NV.

Freeman, Eloise. April 28, 2007. Las Vegas, NV.

Fuller, Tim. July 31, 2007. Las Vegas, NV.

Gable, Gary. July 31, 2007. Henderson, NV.

Garcia, Rio. June 6, 2007. Las Vegas, NV.

Gibson, James. June 12, 2007. Henderson, NV.

Giunchigliani, Chris. June 22, 2007. Las Vegas, NV.

Glennon, Jake. July 2, 2007. Las Vegas, NV.

Goodman, Oscar. June 21, 2007. Las Vegas, NV.

Gordon, Benny. March 20, 2007. Logandale, NV.

Grand, Bernice. July 15, 2005, and February 7, 2007. Las Vegas, NV.

Green, Frank. July 27, 2007. Henderson, NV.

Haynes, Judy. June 6, 2007. Las Vegas, NV.

Henson, JR. February 15, 2007. Las Vegas, NV.

Hossein, Azaan. August 2, 2007. Las Vegas, NV.

Howell, Jeff. July 20, 2007. Las Vegas, NV.

Jameson, Nate. June 4, 2007. Las Vegas, NV.

Joseph, Patricia. June 29, 2007. Las Vegas, NV.

Kaiser, Ian. July 3, 2007. Las Vegas, NV.

Kasciniak, Marty. February 27, 2007. Las Vegas, NV.

Keane, Minnie. July 20, 2007. Las Vegas, NV.

Kubiak, Jon. May 4, 2007. Las Vegas, NV.

Lamm, Rich. June 7, 2007. Las Vegas, NV.

Lewin, Mark. May 22, 2008. Las Vegas, NV.

Lewis, Leonard. June 27, 2007. Las Vegas, NV.

Lewis, Tina. July 17, 2007. Las Vegas, NV.

Lorena. March 20, 2007. Las Vegas, NV.

Lumpkin, Libby. July 17, 2007. Las Vegas, NV.

Lusetto, Rudolf. February 12, 2007. Las Vegas, NV.

Macy. May 23, 2007. Las Vegas, NV.

Mani, Joseph. April 12, 2007. Las Vegas, NV.

Manner, William. August 3, 2007. Las Vegas, NV.

Marlin, Karl. January 30, 2007. Henderson, NV.

Marney. June 8, 2007. Las Vegas, NV.

Martienko, Vera. June 4, 2007. Las Vegas, NV.

Martin, Myron. July 10, 2007. Las Vegas, NV.

Matt. March 24, 2007. Las Vegas, NV.

McAllister, Tom. May 8, 2007. Las Vegas, NV.

McCray, Bob. April 27, 2007. Las Vegas, NV.

Mendez, Marissa. June 6, 2007. Las Vegas, NV.

Merrill, Ralph. July 18, 2007. Las Vegas, NV.

Mitch. December 21, 2005. Las Vegas, NV.

Mont, Jeremy. June 22, 2007. Las Vegas, NV.

Montandon, Mike. May 23, 2007. North Las Vegas, NV.

Moore, Ina. July 8, 2005. Las Vegas, NV.

Morelli, Adam. May 18, 2007. Henderson, NV.

Murakami, Teressa. July 13, 2005. Las Vegas, NV.

Murrell, Ben. May 22, 2007. Las Vegas, NV.

Nakae, Shari. May 29, 2007. Las Vegas, NV.

Nassif, Quadir. July 30, 2007. West Las Vegas, NV.

Nawrocki, Jennifer. March 22, 2007. Las Vegas, NV.

Needham, Sergio. July 19, 2007. North Las Vegas, NV.

Newman, Shawn. March 7, 2007. Summerlin, NV.

Nickel, Peter. June 28, 2007. Henderson, NV.

Okamoto, John. February 13, 2007. Las Vegas, NV.

Park, Lynn. March 22, 2007. Las Vegas, NV.

Pasumarthi, Murali. July 27, 2007. Las Vegas, NV.

Paul, Georgine. June 6, 2007. Las Vegas, NV.

Peters, Sandra. March 3, 2007. Las Vegas, NV.

Pomeroi, Antoine. July 12, 2007. Las Vegas, NV.

Pratt, Bonnie. February 16, 2007. Las Vegas, NV.

Pratt, Orlando. February 16, 2007. Las Vegas, NV.

Randolph, Elijah. July 25, 2007. North Las Vegas, NV.

Randolph, Jonelle. July 25, 2007. North Las Vegas, NV.

Reardon, Glen. July 18, 2007. Las Vegas, NV.

Roberts, Mavie. May 15, 2007. Las Vegas, NV.

Roberts, Melvin. July 26, 2007. West Las Vegas, NV.

Robins, Greg. June 30, 2007. Henderson, NV.

Roe, Michelle. May 31, 2007. Boulder City, NV.

Rothschild, Josef. July 16, 2007. Henderson, NV.

Salvador, Cora. January 16, 2007. Las Vegas, NV.

Salvador, Flint. January 4, 2007. Las Vegas, NV.

Salvador, Jonas. January 16, 2007. Las Vegas, NV.

Sanchez, Armando. August 3, 2007. Las Vegas, NV.

Schwartz, Howard. February 26 and March 20, 2007. Las Vegas, NV.

Seargent, Tad. August 2, 2007. Las Vegas, NV.

Sears, Ian. July 10, 2007. Las Vegas, NV.

Sedillo, Darren. March 10, 2007. Las Vegas, NV.

Sharp, Kent. July 26, 2007. Las Vegas, NV.

Shaw, Renae. March 14 and March 26, 2007. West Las Vegas, NV.

Shaw, Ronald. March 14, 2007. West Las Vegas, NV.

Simmons, Jeff. May 14, 2007. Las Vegas, NV.

Simon, Frank. June 30 and July 19, 2007. Las Vegas, NV.

Simon, Frank, Jr. June 21, 2007. Henderson, NV.

Smith, John L. June 6, 2007. Las Vegas, NV.

Smith, Valerie. July 14, 2007. North Las Vegas, NV.

Snow, Tracy. August 1, 2007. Las Vegas, NV.

Snyder, Donald. July 10, 2007. Las Vegas, NV.

Sparks, Charley. April 11, 2007. Las Vegas, NV.

Stauffer, Jason. July 11, 2007. Las Vegas, NV.

Strong, Devin. March 15, 2007. Las Vegas, NV.

Talin, Amy. July 27, 2007. Las Vegas, NV.

Taylor, Hank. June 11, 2007. Las Vegas, NV.

Toomey, George. August 1, 2007. Las Vegas, NV.

Torrella, Darrell. June 7, 2007. Summerlin, NV.

Trimble, Don. February 27 and May 24, 2007. Las Vegas, NV.

Walker, Aric. March 8 and May 16, 2007. Las Vegas, NV.

Wilkes, Lance. January 5, 2007. Henderson, NV.

Wychof, Ben. January 31, 2007. Las Vegas, NV.

Zanelli, Al. July 14, 2007. Las Vegas, NV.

OTHER SOURCES

Airriess, Christopher A., Wei Li, Karen J. Leong, Angela Chia-Chen Chen, and Verna M. Keith. "Church-Based Social Capital, Networks, and Geographical Scale: Katrina Evacuation, Relocation, and Recovery in a New Orleans Vietnamese American Community." *Geoforum* 39, no. 3 (2008): 1333–46.

American Association of State Highway and Transportation Officials (AASHTO). *Transportation: Invest in Our Future—America's Freight Challenge.* Washington DC: AASHTO, 2007.

Basso, Keith H. "Wisdom Sits in Places: Notes on a Western Apache Landscape." In *Senses of Place,* edited by Steven Feld and Keith H. Basso, 53–90. Santa Fe: School of American Research Press, 1996.

Bauer, John. "Stability and Change in United States Religious Regions, 1980–2000." PhD diss., University of Kansas, 2006.

———. "U.S. Religious Regions Revisited." *Professional Geographer* DOI:10.1080/003301 24.2011.611429 (2011).

Beville, John M. "How Las Vegas Pioneered a Rotary Club." *Nevada Historical Society Quarterly* 10, no. 3 (1967): 29–34.

Borchard, Kurt. *Homeless in Las Vegas: Stories from the Street.* Reno: University of Nevada Press, 2011.

———. *The Word on the Street: Homeless Men in Las Vegas.* Reno: University of Nevada Press, 2005.

Brents, Barbara G., Crystal A. Jackson, and Kathryn Hausbeck, *The State of Sex: Tourism, Sex and Sin in the New American Heartland.* New York: Routledge, 2010.

Burchell, Robert W., Anthony Downs, Barbara McCann, and Sahan Mukherji. *Sprawl Costs: Economic Impacts of Unchecked Development.* Washington DC: Island Press, 2005.

Casey, Edward S. "How to Get from Space to Place in a Fairly Short Stretch of Time: Phenomenological Prolegomena." In *Senses of Place,* edited by Steven Feld and Keith H. Basso, 13–52. Santa Fe: School of American Research, 1996.

Center for Business and Economic Research. *Southern Nevada Factbook,* Spring 2006. University of Nevada, Las Vegas, 2006.

Cooper, Marc. *The Last Honest Place in America: Paradise and Perdition in the New Las Vegas.* New York: Nation Books, 2004.

Corbett, Julia Mitchell. "Religion in the United States: Notes toward a New Classification." *Religion and American Culture* 3, no. 1 (1993): 91–112.

De Wit, Cary. "Field Methods for Investigating Sense of Place." *North American Geographer* 5, no. 1–2 (2003): 5–30.

———. "Women's Sense of Place on the American High Plains." *Great Plains Quarterly* 21 (2001): 29–44.

Dunbar, Andrew J., and Dennis McBride. *Building Hoover Dam: An Oral History of the Great Depression*. New York: Twayne, 1993.

Elliott, Gary E. *The New Western Frontier: An Illustrated History of Greater Las Vegas*. Carlsbad, CA: Heritage Media Corp., 1999.

Ernst, Michelle, Marisa Lang, and Stephen Davis. *Dangerous by Design*. Washington DC: Transportation for America.

Feld, Steven, and Keith H. Basso, eds. *Senses of Place*. Santa Fe: School of American Research Press, 1996.

Fiero, Bill. *Geology of the Great Basin*. Reno: University of Nevada Press, 1986.

Findlay, John M. *People of Chance: Gambling in American Society from Jamestown to Las Vegas*. New York: Oxford University Press, 1986.

Foote, Kenneth E. *Shadowed Ground: America's Landscapes of Violence and Tragedy*. Rev. ed. Austin: University of Texas Press, 2003.

Frazier, Ian. *Great Plains*. New York: Farrar, Straus and Giroux, 1989.

Frehner, Brian. "'Squeezing the Juice out of Las Vegas': Reflections on Growing Up in Smalltown, USA." In *The Grit Beneath the Glitter: Tales from the Real Las Vegas*, edited by Hal K. Rothman and Mike Davis, 187–94. Berkeley and Los Angeles: University of California Press, 2002.

Futrell, Robert, Cristie Batson, Barbara G. Brents, Andrew Dassopoulos, Crissy Nicholas, Mark J. Salvaggio, and Candace Griffith. *Las Vegas Metropolitan Area Social Survey: 2010 Highlights*. Las Vegas: Department of Sociology, University of Nevada, Las Vegas, 2010.

Garreau, Joel. *Edge City: Life on the New Frontier*. New York: Doubleday, 1991.

Geran, Trish. *Beyond the Glimmering Lights: The Pride and Perseverance of African Americans in Las Vegas*. Las Vegas: Stevens Press, 2006.

Goin, Peter. "Visual Literacy." *Geographical Review* 91, no. 1–2 (2001): 363–69.

Goodwin, Joanne L. "'She Works Hard for Her Money': A Reassessment of Las Vegas Women Workers, 1945–1985." In *The Grit Beneath the Glitter: Tales from the Real Las Vegas*, edited by Hal K. Rothman and Mike Davis, 243–59. Berkeley and Los Angeles: University of California Press, 2002.

Gorelow, Andrew S., and Paul H. Skrbac. *Climate of Las Vegas, Nevada*. Las Vegas: National Weather Service, 2005.

Gottdiener, Mark, Claudia C. Collins, and David R. Dickens. *Las Vegas: The Social Production of an All-American City*. Malden, MA: Blackwell, 1999.

Hansen, Maia. "Skin City." In *The Real Las Vegas: Life Beyond the Strip*, edited by David Littlejohn, 217–41. New York: Oxford University Press, 1999.

Hausbeck, Kathryn. "Who Puts the 'Sin' in 'Sin City' Stories? Girls of Grit and Glitter in the City of Women." In *The Grit Beneath the Glitter: Tales from the Real Las Vegas*, edited by Hal K. Rothman and Mike Davis, 335–46. Berkeley and Los Angeles: University of California Press, 2002.

Herbert, Steve. "For Ethnography." *Progress in Human Geography* 24, no. 4 (2000): 550–68.

Hicks, Terri Lynn. "Quality of Life Amidst the Lights of Las Vegas." Master's thesis, University of Nevada, Las Vegas, 1999.

Hopkins, A. D., and K. J. Evans, eds. *The First 100: Portraits of the Men and Women Who Shaped Las Vegas.* Las Vegas: Huntington Press, 1999.

Jackson, J. B. *Discovering the Vernacular Landscape.* New Haven, CT: Yale University Press, 1984.

———. *A Sense of Place, a Sense of Time.* New Haven, CT: Yale University Press, 1994.

———. "The Stranger's Path." In *Landscapes: Selected Writings of J. B. Jackson,* edited by Ervin H. Zube, 92–106. Amherst: University of Massachusetts Press, 1970.

Jacobs, Jane. *The Death and Life of Great American Cities.* New York: Random House, 1961.

Jakle, John A. *The Visual Elements of Landscape.* Amherst: University of Massachusetts Press, 1987.

Land, Barbara, and Myrick Land. *A Short History of Las Vegas.* Reno: University of Nevada Press, 1999.

Leavitt, Phyllis M. "Depression Pioneers of Boulder City." In *Nevada Official Bicentennial Book,* edited by Stanley W. Paher, 286–89. Las Vegas: Nevada Publications, 1976.

Leibovich, Lori. "Houses of the Holy." In *The Real Las Vegas: Life Beyond the Strip,* edited by David Littlejohn, 167–79. New York: Oxford University Press, 1999.

Lewis, Peirce F. *New Orleans: The Making of an Urban Landscape.* 2nd ed. Santa Fe: Center for American Places, 2003.

Littlejohn, David, ed. *The Real Las Vegas: Life Beyond the Strip.* New York: Oxford University Press, 1999.

Longwell, C. R., E. H. Pampeyan, B. Bowyer, and R. J. Roberts. *Geology and Mineral Deposits of Clark County, Nevada: Nevada Bureau of Mines and Geology, Bulletin 62.* Reno: Mackay School of Mines, University of Nevada in Cooperation with the United States Geological Survey, 1965.

McMackin, Shannon. "I Didn't Know Anybody Lived There." In *The Grit Beneath the Glitter: Tales from the Real Las Vegas,* edited by Hal K. Rothman and Mike Davis, 195–207. Berkeley and Los Angeles: University of California Press, 2002.

Meinig, Donald W. "Geography as an Art." *Transactions of the Institute of British Geographers* 8, no. 3 (1983): 314–28.

———. "Symbolic Landscapes: Some Idealizations of American Communities." In *The Interpretation of Ordinary Landscapes,* ed. Donald W. Meinig, 164–92. New York: Oxford University Press, 1979.

Moehring, Eugene P. "Las Vegas History: A Research Agenda." *Nevada Historical Society Quarterly* 47, no. 4 (2004): 259–82.

———. "Public Works and the New Deal in Las Vegas, 1933–1940." *Nevada Historical Society Quarterly* 24, no. 2 (1981): 107–29.

———. *Resort City in the Sunbelt: Las Vegas 1930–2000.* 2nd ed. Reno: University of Nevada Press, 2000.

Moehring, Eugene P., and Michael S. Green. *Las Vegas: A Centennial History.* Reno: University of Nevada Press, 2005.

Moskowitz, Lisa. "For Sale." In *The Real Las Vegas: Life Beyond the Strip,* edited by David Littlejohn, 147–66. New York: Oxford University Press, 1999.

National Gambling Impact Study Commission. *Final Report.* Washington DC: National Gambling Impact Study Commission, 1999.

Nevada Chapter of the American Institute of Architects. *Blueprint for Nevada.* Las Vegas: AIA Nevada, 2008.

Nystrom, Eric. "Labor Strife in Las Vegas: The Union Pacific Shopmen's Strike of 1922." *Nevada Historical Society Quarterly* 44, no. 4 (2001): 313–32.

O'Brien, Matthew. *Beneath the Neon: Life and Death in the Tunnels of Las Vegas.* Las Vegas: Huntington Press, 2007.

Paher, Stanley W. *Las Vegas, as It Began—as It Grew.* Las Vegas: Nevada Publications, 1971.

Parker, Robert E. "The Social Costs of Rapid Urbanization in Southern Nevada." In *The Grit Beneath the Glitter: Tales from the Real Las Vegas,* edited by Hal K. Rothman and Mike Davis, 126–44. Berkeley: University of California Press, 2002.

Pocock, D.C.D. "Place and the Novelist." *Transactions of the Institute of British Geographers* 6, no. 3 (1981): 337–47.

Rafferty, Kevin A. *Cultural Resources Overview of the Las Vegas Valley, Technical Report No. 13.* Reno: Bureau of Land Management, 1984.

Regional Transportation Commission of Southern Nevada. *Regional Transportation Plan: 2009–2030.* Las Vegas, NV: RTC

Relph, Edward. *Place and Placelessness.* London: Pion Limited, 1976.

Riley, Robert R. "The Visible, the Visual, and the Vicarious: Questions About Vision, Landscape, and Experience." In *Understanding Ordinary Landscapes,* edited by Paul Groth and Todd W. Bressi, 200–209. New Haven, CT: Yale University Press, 1997.

Roske, Ralph J. *Las Vegas: A Desert Paradise.* Tulsa, OK: Continental Heritage Press, 1986.

Rothman, Hal. *Neon Metropolis: How Las Vegas Started the Twenty-First Century.* New York: Routledge, 2002.

Rothman, Hal K., and Mike Davis, eds. *The Grit Beneath the Glitter: Tales from the Real Las Vegas.* Berkeley and Los Angeles: University of California Press, 2002.

Rowley, Rex J. "Becoming Las Vegas: Opportunity and Challenge with the Building of Boulder Dam." *Nevada Historical Society Quarterly,* forthcoming.

Ryden, Kent C. *Mapping the Invisible Landscape: Folklore, Writing, and the Sense of Place.* Iowa City: University of Iowa Press, 1993.

Sanchez, Marie. "Growing Up in Las Vegas." In *The Real Las Vegas: Life Beyond the Strip,* edited by David Littlejohn, 75–96. New York: Oxford University Press, 1999.

Schein, Richard. "The Place of Landscape: A Conceptual Framework for Interpreting an American Scene," *Annals of the Association of American Geographers* 87, no. 4 (1997): 660–80.

Schnell, Steven M. "Creating Narratives of Place and Identity in 'Little Sweden, U.S.A.'" *Geographical Review* 93, no. 1 (2003): 1–29.

Schrank, David, Tim Lomax, and Shawn Turner. *Urban Mobility Report 2010*. College Station: Texas Transportation Institute, 2010.

Schumacher, Geoff. *Sun, Sin, and Suburbia: An Essential History of Modern Las Vegas*. Las Vegas: Stephens Press, 2004.

Schwartz, David G. *Suburban Xanadu: The Casino Resort on the Las Vegas Strip and Beyond*. Routledge: New York, 2003.

Schwer, R. Keith, Sandra Phillips Johnson, Rennae Daneshvary, Richard W. Hoyt, and Saadullah Bashir. *2007 Southern Nevada Community Assessment*. Las Vegas: Center for Business and Economic Research, University of Nevada, Las Vegas/United Way of Southern Nevada, 2007.

Shepperson, Wilbur S., ed. *East of Eden, West of Zion: Essays on Nevada*. Reno: University of Nevada Press, 1989.

Simich, Jerry L., and Thomas C. Wright, eds. *The Peoples of Las Vegas: One City, Many Faces*. Reno: University of Nevada Press, 2005.

Smith, John L., and Patricia Smith. *Moving to Las Vegas*. 3rd ed. Fort Lee, NJ: Barricade Books, 2003.

Smith, Ronald W., ed. *The Las Vegas Metropolitan Area Project: Studies of Selected Quality of Life Issues*. Las Vegas: University of Nevada, Las Vegas, 1998.

Soil Conservation Service. *Soil Survey of Las Vegas Valley Area Nevada, Part of Clark County*. Washington DC: US Department of Agriculture, 1985.

Steensland, Brian, Jerry Z. Park, Mark D. Regnerus, Lynn D. Robinson, W. Bradford Wilcox, and Robert D. Woodberry. "The Measure of American Religion: Toward Improving the State of the Art." *Social Forces* 79, no. 1 (2000): 291–318.

Stegner, Wallace. *Where the Bluebird Sings to the Lemonade Springs: Living and Writing in the West*. New York: Random House, 1992.

Stevens, Joseph E. *Hoover Dam: An American Adventure*. Norman: University of Oklahoma Press, 1988.

Thompson, William N., and R. Keith Schwer. "Beyond the Limits of Recreation: Social Costs of Gambling in Southern Nevada." *Journal of Public Budgeting, Accounting, and Financial Management* 17, no. 1 (2005): 62–93.

Tuan, Yi-Fu. *Space and Place: The Perspective of Experience*. Minneapolis: University of Minnesota Press, 1977.

Volberg, Rachel A. "Gambling and Problem Gambling Among Adolescents in Nevada: Report to the Nevada Department of Human Resources." North Hampton, MA: Gemini Research, 2002.

———. "Gambling and Problem Gambling in Nevada: Report to the Nevada Department of Human Resources." North Hampton, MA: Gemini Research, 2002.

Ward, Kenric F. *Saints in Babylon: Mormons and Las Vegas*. Self-published, 2002.

Western, John. *A Passage to England: Barbadian Londoners Speak of Home*. Minneapolis: University of Minnesota Press, 1992.

Wiley, Peter, and Robert Gottlieb. *Empires in the Sun: The Rise of the New American West*. New York: G. P. Putnam's Sons, 1982.

Wilhite, Jud, with Bill Taaffe. *Stripped: Uncensored Grace on the Streets of Vegas.* Colorado Springs, CO: Multnomah Books, 2006.

Wright, John K. "Terrae Incognitae: The Place of Imagination in Geography." *Annals of the Association of American Geographers* 37, no. 1 (1947): 1–15.

Zelinsky, Wilbur. *The Cultural Geography of the United States.* Englewood Cliffs, NJ: Prentice-Hall, 1973.

INDEX